RACE IN MIND

Race, IQ, and Other Racisms

Alexander Alland, Jr.

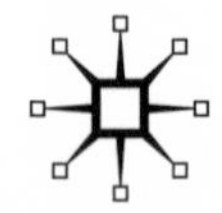

RACE IN MIND

First published 2002 by
PALGRAVE MACMILLAN™
175 Fifth Avenue, New York, N.Y. 10010 and
Houndmills, Basingstoke, Hampshire, England RG21 6XS.
Companies and representatives throughout the world.

PALGRAVE MACMILLAN IS THE GLOBAL ACADEMIC IMPRINT OF THE PALGRAVE MACMILLAN
division of St. Martin's Press, LLC and of Palgrave Macmillan Ltd.
Macmillan® is a registered trademark in the United States, United Kingdom and other countries. Palgrave is a registered trademark in the European Union and other countries.

ISBN 0–312–23838-X hardback

Library of Congress Cataloging-in-Publication Data
Alland, Alexander, 1931-
Race in mind : race, IQ, and other racisms / by Alexander Alland, Jr.
p. cm.
Includes bibliographical references and index.
ISBN 0–312–23838-X
1. Race. 2. Intelligence levels. 3. Racism—United States—History. I. Title.

HT1521.A39 2002
305.8—dc21

2002020117

A catalogue record for this book is available from the British Library.

Design by Letra Libre, Inc.

First edition: September 2002
10 9 8 7 6 5 4 3 2

Printed in the United States of America

CONTENTS

This book is dedicated to my son and daughter-in-law, David and Leila Alland, their children, Jonathan and Isabel, my daughter and son-in-law, Julie Alland and Alan Hopkins, and to my former student from whom I have learned so much, Joel Wallman.

Thanks are due to my editor at Palgrave Macmillan, Kristi Long, and my incredibly efficient copyeditors and proofreaders, Jen Simington and Annje Kern, as well as my friend and colleague at Columbia University, Jill Shapiro. They helped to make this a better book. All errors are, of course, my own.

CHAPTER ONE

PROLOGUE

WHY I HAVE WRITTEN THIS BOOK AND WHY I THINK IT IS IMPORTANT

Let me begin by quoting from an original document in my possession that eloquently if inadvertently illustrates the horrors of slavery in the United States. It is a flyer advertising a "Public Auction of Negroes to be held at Potters Mart in Charleston, South Carolina on March 5th, 1833."

> Conditions: 1/2 cash, balance by bond, bearing interest from date of sale. Payable in one to two years to be secured by a mortgage of the Negroes, and appraised personal security. Auctioneer will pay for the papers.
>
> A valuable Negro woman, accustomed to all kinds of house work. Is a good plain cook, and excellent dairy maid, washes and irons. She has four children, one a girl about 13 years of age, another 7, a boy about 5, and an infant 11 months old. Two of the children will be sold with the mother, the other separately, if it best suits the purchaser.
>
> A very valuable Blacksmith, wife and daughters; the Smith is in the prime of life, and a perfect master of his trade. His wife about 27 years old, and his daughters 12 and 10 years old have been brought up as house servants, and as such are very valuable. Also for sale 2 likely young negro wenches, one of whom is 16 the other 13, both of whom have been taught and accustomed to the duties of house servants. The 16 year old wench has one eye.

> A likely yellow girl about 17 years old, has been accustomed to all kinds of house and garden work. She is sold for no fault. Sound as a dollar.
>
> House servants: The owner of a family described herein would sell them for a good price only, they are offered for no fault whatever, but because they can be done without, and money is needed. He has been offered $1250. They consist of a man 30 to 33 years old, who has been raised in a genteel Virginia family as house servant, Carriage driver etc., in all which he excels. His wife a likely wench of 25 to 30 raised in like manner, as chamber maid, seamstress, nurse etc., their two children, girls of 12 and 4 or 5. They are bright mulattoes, of mild tractable dispositions, unassuming manners and of genteel appearance and well worthy the notice of a gentleman of fortune.
>
> Also 14 Negro Wenches ranging from 16 to 25 years of age, all sound and capable of doing a good day's work in the house or field.

Worse! Joyce Carol Oates, writing in the June 21, 2001, issue of the *New York Review of Books,* quotes the following ad from the *Natchez Free Trader:*

> TO PLANTERS AND OTHERS.—
>
> *Wanted fifty Negroes.* Any person having *sick negroes,* considered *incurable* by their respective physicians, . . . and wishing to dispose of them, Dr. Stillman will pay cash for Negroes affected with scrofula or king's evil, confirmed hypochondriacism, apoplexy, or diseases of the brain, kidneys, spleen, stomach, and intestines, bladder, and its appendages, diarrorea, dysentery, etc. *The highest cash price will be paid as above.*

Oates goes on to tell the reader that the advertiser was a medical school instructor searching for specimens to dissect. She writes "They keep [afflicted slaves] on hand, and when they need one they bleed him to death."

These are two examples of many that illustrate how blacks were considered mere property, even to the point where children could be sold away from their parents and the sick turned into specimens. Reading these one can well understand the development of the abolitionist movement among those people of conscience who felt that slavery was an immoral blot on a nation that took pride in its young democracy. But we cannot assume that even the majority of abolitionists at the time believed that Blacks were the intellectual equals of whites.

Thomas Henry Huxley, an English biologist known as "Darwin's Bulldog" due to his energetic advocacy of Darwin's theory, wrote the following in an article, "Emancipation—Black and White," at the close of the Civil War in the United States.

> Quashie's plaintive inquiry, "Am I not a man and a brother?" seems at last to have received its final reply—the recent decision of the fierce trial by battle on the other side of the Atlantic fully concurring with that long since delivered here in a more peaceful way.
>
> The question is settled; but even those who are most thoroughly convinced that the doom is just, must see good grounds for repudiating half the arguments which have been employed by the winning side; and for doubting whether its ultimate results will embody the hopes of the victors, though they may more than realize the fears of the vanquished. It may be quite true that some Negroes are better than some white men; but no rational man, cognizant of the facts, believes that the average negro is the equal, still less the superior, of the average white man. And, if this be true, it is simply incredible that, when all his disabilities are removed, and our prognathous relative has a fair field and no favor, as well as no oppressor, he will be able to compete successfully with his bigger-brained and smaller-jawed rival, in a contest to be carried on by thoughts and not by bites. The highest places in the hierarchy of civilization will assuredly not be within the reach of our dusky cousins, though it is by no means necessary that they should be restricted to the lowest.
>
> But whatever the position of stable equilibrium into which the laws of social gravitation may bring the negro, all responsibility for the result will henceforward lie between nature and him. The white man may wash his hands of it, and the Caucasian conscience be void of reproach for evermore. And this, if we look to the bottom of the matter, is the real justification for the abolition policy. (Huxley 1910, 115)

Huxley's conclusions, written from England, concerning the biological effects of abolition in the United States were echoed in our country by Lewis Henry Morgan, a prominent Rochester lawyer, member of the New York State Legislature, and major figure in early American anthropology.

Although an abolitionist, Morgan felt strongly that left to their own devices, blacks in the United States would disappear due to their inferior biological status, which he, like so many others, took as a given of the evolutionary process.

Well after emancipation, average Americans, white and black, were exposed, via the mass media, to a depressingly frequent degrading depiction of blacks. For an illustration of this, let us fast-forward to 1937. In the August 9 issue of that year, *Life* magazine published a cover story under the title "Watermelons to Market." At the time, and well into the 1950s, the mass circulation weekly *Life* was *the* major source of news photography in the United States. The cover of this issue carries a full-page photograph of a black man, his back to the photographer, perched on a tractor loaded with watermelons. The article begins on page 51. The text is framed, top and bottom, with photos of the harvesting and marketing procedures. The next page shows three photos under the caption "All Southerners Like Watermelon." The first photo is of young white women on a beach. It bears the caption: "A watermelon picnic is considered the height of fun by Georgia girls, who like to cool their melons in a creek. The girl seated at right is going to get sand in hers." The second photo shows a black woman nursing her baby at her bare breast as she holds a large piece of watermelon up to her face in her left hand. The caption reads: "Nothing makes a Negro's mouth water like a luscious, fresh-picked melon. Any colored 'mammy' can hold a huge slice in one hand while holding her offspring in the other. Since the watermelon is 92% water, tremendous quantities can be eaten." The last photo is of a herd of pigs eating watermelon. The caption reads: "What melons the Negroes do not consume will find favor with the pigs."

School segregation in the South, finally outlawed by a Supreme Court decision in 1954, was closely linked to the notion that blacks were inferior in intelligence to whites. When segregation was overturned, many in the South as well as in the North protested, claiming that integration would do irreparable harm to white children. One of the most outspoken northerners to take issue with the court's decision was Carlton Putnam, Columbia Law School graduate, former chairman of Delta Airlines, and a biographer of Theodore Roosevelt. In 1958 Mr. Putnam, responding to a *Life* magazine editorial supporting the decision (*Life* had changed its politics considerably by the 1950s), wrote a letter to President Eisenhower defending segregation that was also published in a Memphis newspaper, the *Commercial Appeal.* This was followed by a longer letter published in the *Richmond Times-Democrat.* This second letter drew a great deal of attention from readers, and Putnam asked the editors to send him copies that criticized his position. His response to these letters was then used in Putnam's book, *Race and Reason: A Yankee View,* published by the Public Affairs Press in Washington, D.C., in 1961:

> White Southerners understood the Negro and in large measure loved him. . . . The South, after generations of experience, had developed customs and a way of life with the Negro that took his limitations into consideration with a minimum of friction and a maximum of kindness. (Putnam 1961, 21)
>
> There is no basis in sound science for the assumption, promoted by various minority groups in recent decades, that all races are biologically equal in their capacity for advance, or even to sustain, what is commonly called Western civilization. (Putnam 1961, 36)

In respect to the notion that there are intellectually successful blacks occupying important positions in American society, Putnam replies: "Let me point out that not only are these Negroes in no sense typical of their race, whose genes they nevertheless carry and will pass on to their children, but that most of them owe their ability to some percentage of white genes in the system" (Putnam 1961, 36).

Carlton Putnam died at the age of 61 in the year 1998. In an obituary published in the *New York Times* of March 16, 1998, Robert McG. Thomas Jr. says of Putnam's racist ideas:

> Mr. Putnam devoted much of his book to arguing that when it came to the personal characteristics that produced the glories of Western civilization, the Negro race could not hold a candle to the white race. The evidence he amassed was so impressive and so thoughtfully presented, it was easy to overlook the fact that it was irrelevant. The Supreme Court, after all, had not used sociological evidence to establish that black people were the intellectual equals of white people but only that they had been harmed by forced segregation.

Astonished by these words I wrote a letter to the *Times* (unpublished) in which I noted that while it might seem cruel to criticize an obituary I could not let its author get away with saying that Carlton Putnam's ideas concerning race were either impressive or thoughtfully presented. As evidence I included the first two of the quotes from Putnam's book presented above!

The question must be asked: Was it so astonishing that the *Times* let these notions go by without any disclaimer on their part? I think the answer must be "no" since every time a book is published concerning the assumed link between race and intelligence it is featured, most often in a positive light, by even the most serious of the popular press. While the *Times* did publish a negative review of the *Bell Curve* in its daily edition, the first notice of the book in the *Times* came in the widely read *Sunday Book Review* section. There it was treated primarily as a serious work of science in a long article that also lent credence to another highly flawed work by the Canadian psychologist J. Philippe Rushton (to be discussed later in chapter nine). Unfortunately, a wide segment of the American public, including government officials, lies in wait for such confirmation of their prejudices concerning race in general and race and IQ in particular.

This is the reason why, though only a small minority of scholars in the English-speaking world makes claims concerning a strong correlation between race and IQ, their influence goes well beyond their numbers. These scholars attract immediate attention as they vigorously support the argument that many of our educational problems, particularly the system's failures, have been caused by the refusal to accept the fact that intelligence, as measured on IQ tests, is primarily genetic in origin and, more importantly, that IQ is unequally distributed by race in the general population. Most of these scholars have attempted to prove that the average black is inferior to the average white. Some of these same scholars, perhaps in an attempt to show that they are not racists, also claim that the average Asian is superior to the average white, but only by a few IQ points. Additionally, although this issue will not be treated in this volume, some of these scholars believe that the average woman, although she may have some inherited skills superior to those of men, is inferior to the average male in intelligence. Suffice it to say here that the beliefs held by these individuals about women are congruent with certain beliefs concerning blacks and some other minorities.

Although there are professed racists in the United States and elsewhere, few of the academics who deal with the problem of race and genetics care to place themselves in this category. It is their common strategy to protest that it is their enemies who so label them in an unscholarly attempt to discredit their ideas.

The two academic subjects that have been most concerned with what I shall refer to as the "IQ argument" are anthropology and psychology. While the vast majority of contemporary anthropologists reject the concept of race out of hand and are, to say the least, skeptical when it comes to relationships between any form of population typing and IQ, many psychologists still cling to both notions.

While this book concentrates on scholars who have argued that race and IQ are intimately linked because the issue of race goes well beyond the notion of IQ (see, for example, the articles published in the *New York Times* in six issues during the month of June 2000; see Lelyveld 2001) and has a major negative impact on American culture, I expand my discussion of the race issue in the last chapter to deal with some of the other anthropological and sociological issues concerning race that are current in contemporary U.S. life. I have written this book because I am convinced that the misguided concept of race as biologically based continues to be dangerous not only for minority groups who are the targets of racism but also for the well being of American society as a whole.

This book takes an extremely critical view of the authors about to be discussed. The reader, therefore, may be curious about how, if their works are so flawed, they manage to find significant financial support for their work. The mystery can be cleared up easily, however. First of all, every author discussed, with the exception of Robert Ardrey, had or still has a tenured post as full professor. Both William Shockley and Konrad Lorenz were Nobel laureates, but, I hasten to add, Shockley, a physicist, won his for having invented the transistor, a fact that has nothing to do with competence in genetics, psychology, or anthropology. Lorenz's research was, throughout his career, limited to nonhuman organisms, primarily birds and fish. Neither Ardrey (who was a professional playwright) nor Lorenz had any expertise in the area of human behavior, not to mention race or IQ. Carlton S. Coon was a tenured professor of physical anthropology at Harvard. His notions about the evolution of races are not supported by any of his own research. Richard J. Herrnstein was, until his death, a tenured professor of psychology at Harvard, where he was noted for his work in psychological statistics rather than intelligence. Charles Murray holds a Ph.D. in political science and works for a conservative think tank in Washington, D.C. This foundation is not noted for racist positions but is heavily invested in the idea that successful individuals in high places get there primarily by virtue of their own intelligence and hard work. Such an elitist point of view is known as "meritocracy." Michael E. Levin is a tenured professor of philosophy at the City College of the city of New York. He keeps his racial views out of the classroom, limiting them to publication in what I will argue are rather dubious journals. Leonard Jeffries is the former head of the black studies department and currently a tenured professor at the City University of New York. There is no mystery concerning Hans Eysenck, Arthur R. Jensen, and J. Philippe Rushton. Jensen studied with Eysenck, and Rushton with Jensen. So, in a real sense they constitute a genealogy in respect to their basic ideas. Financially all three have received heavy support from the Pioneer Fund, a private agency noted for its support of racist research.

In his book, *The Nazi Connection: Eugenics, American Racism, and German National Socialism,* Steffan Kuhl, notes that the founders of the Pioneer Fund had praised aspects of Nazi Germany's racial policies and had given financial support to controversial research into race and IQ.

Although the fund's money came from the New England millionaire Wickliffe Draper, in its formative years it was run by Harry H. Laughlin and Fredrick Osborn, both eugenicists.

> Scientists who played a leading role in the American eugenics movement, and, as I will illustrate, who supported Hitler's race policy, initiated the Pioneer Fund in 1937. The Fund's stated purpose was to "improve the character of the American people" by encouraging the procreation of descendants of "white people who settled in the original thirteen colonies prior to the adoption of the constitution and/or from related stocks" and to

> provide aid in conducting research on "race betterment with special reference to the people of the United States." (Kuhl 1994, 5–6)

In an article published in the *Hamilton (Ontario) Spectator* for April 17, 2000, Steve Buist, discussing J. Philippe Rushton's research funding from the Pioneer Fund, has the following to say:

> According to the original charter of the Pioneer Fund, the organization would support research that was directed toward "race betterment," with special consideration given to scholarship programs aimed at "children who are deemed to be descended predominantly from white persons who settled in the original 13 States."
>
> Henry Garrett a Pioneer Fund director during the 1970s, organized an international group of scholars dedicated to preventing the mixing of races, preserving segregation and promoting the principle of what was described as "race hygiene."

The reader needs to be alerted to another suspicious aspect of the race-IQ argument. In recent years a great deal of attention, particularly in the press, but also in professional circles within anthropology and psychology, has focused on the new field of sociobiology, now known, at least within psychology, as evolutionary psychology. First, let me make it clear that while I am skeptical of a good deal of the research carried out in this field when it is applied to what are seen as cultural *differences* among human populations I do not deny that our species, like all others, is the product of biological evolution. We share the capacity for cultural behavior and language as products of that evolution. I am convinced, however, that cultural *differences* are the product of our capacity to change our behavioral patterns rapidly precisely because they are *not* linked to our genes. While I must also stress that most sociobiologists avoid the race-IQ argument, some have used it to explain why they believe certain races are inferior. This is the case for Konrad Lorenz and his popularizer, Robert Ardrey, who in his last book on the subject, *The Social Contract,* argues, if somewhat guardedly, for a link between race and intelligence. Lorenz himself, during and after the Nazi period, published ideas that must be categorized as racist. As the reader will see in more detail later, Lorenz believed that inborn aesthetic preferences in mate choice fostered inbreeding within "pure" racial groups and were weakened by civilization. This modern tendency produces "mongrelization" in formerly racially pure populations, leading to inferior individuals. A sociobiological argument is also used by J. Philippe Rushton in order to explain what he believes is low IQ in blacks in comparison with whites. The difference is attributed to different adaptational breeding and nurturance patterns in the two "races." A criticism of this approach appears in chapter nine of this book.

Unfortunately the race-IQ argument remains popular in spite of advances in population biology that disprove the existence of biological race among humans. The race concept is so strong that sometimes, in moments of devilish reverie (male menopause?), I think of it as an easy way to achieve instant notoriety and, perhaps, considerable financial reward. All I need do is publish a book renouncing the views on race and IQ that I have been publicly advocating for the last thirty years: that the term "race" has no place in proper biological thinking about our species and that IQ has nothing to do with so-called racial identity, which has little if anything to do with the complex notion of intelligence (or better—intelligences). Such a change of heart would be "newsworthy" if only because large segments of the public are so willing to accept simple biological explanations for what are actually complicated social problems. This is exactly what happened in 1968, when Arthur R. Jensen published his report claiming that Project Head Start was unlikely to help ghetto children raise their IQs or school performance because he, Jensen, had proof that IQ was at least 80 percent genetic. Although Jensen's data were highly flawed (see chapter five) the "news" made the national press within one week and opened a polemic that has persisted ever since.

Now, I have no intention of changing my views. I present the above scenario to underline the fact that "scientific" racism is a pernicious phenomenon that demands constant vigilance and strongly reasoned counter arguments. This book will offer such arguments. It is written as a response to "scientific" racism that has come for the most part from the English-speaking academic community since the middle of this century. Examples of it turn up at a relatively constant rate, about once every generation, in academic publications as well as the press, both intellectual and popular. The most recent manifestation appears in two books: *The Bell Curve* by Richard J. Herrnstein and Charles Murray (The Free Press,

1994) and *Race, Evolution and Behavior: A Life History Perspective* by J. Philippe Rushton (Transaction Publishers, 1995). The latter author, with the continuing help of the Pioneer Fund, sent 35,000 unsolicited copies of an abridged version of this work to psychologists, sociologists, and anthropologists in Canada and the United States. A second edition of the abridged work was sent out, again unsolicited, to scholars again in 2000.

Herrnstein and Murray's book is an intimidating work of 845 pages replete with graphs and charts. The unwary lay reader may easily accept it as scholarly proof that IQ and race are genetically linked. This is unfortunate since the work is flawed in its most basic conception, ignoring as it does major problems concerning the application of data on the genetic contribution of IQ in one population to another different population.

As is so often the case when IQ and race are the issues, Herrnstein and Murray are disingenuous in making claims to objectivity. They *do* say "It seems highly likely to us that both genes and the environment have something to do with racial differences. What might the mix be? *We are resolutely agnostic on this issue; as far as we can determine, the evidence does not yet justify an estimate*" (311, italics mine). But *The Bell Curve* is loaded with rather uncritical arguments in favor of significant genetically based differences in IQ among the races, while studies that support the environmental argument are downplayed.

With few exceptions, the authors discussed in this book protest their innocence. They accuse their critics of favoring censorship, of being unreconstructed leftists. They employ what I call a "reverse *ad hominem*" expressing a peculiar kind of "*mea culpa*—well sort of" as they profess sadness over the truth about IQ and race. The authors fervently claim their work should not be taken as racist but rather as an urgent call to save all of us, black and white alike, from the dangers posed to society by a concentration of high and low IQs in different segments of the population. Their fondest wish is the improvement of social conditions for all people everywhere. Meanwhile, they cry all the way to the bank.

Arthur R. Jensen concludes his 1972 book, *Genetics and Education,* with the following disclaimer:

> We must clearly distinguish between research on racial differences and racism. Racism implies hate or aversion and aims at denying equal rights

> and opportunities to persons because of their racial origin. It should be attacked by enacting and enforcing laws and arrangements that help to insure equality of civil and political rights and to guard against racial discrimination in educational and occupational opportunities. But to fear research on genetic racial differences in abilities, is, in a sense, to grant the racists assumption: that if it should be established beyond reasonable doubt that there are biological or genetically conditioned differences in mental abilities among individuals or groups, then we are justified in oppressing or exploiting those who are most limited in genetic endowment. This is, of course, a complete *non sequitur.*
>
> I have always advocated dealing with persons as individuals, and I am opposed to according differential treatment to persons on the basis of their race, color, national origin, or social class background. But I am also opposed to ignoring or refusing to investigate the causes of the well-established differences among racial groups in the distribution of educationally relevant traits, particularly IQ. Purely environmental explanations of racial differences in intelligence will never gain the status of scientific knowledge unless genetic theories are put to the test and disproved by the evidence. (Jensen 1972, 329)

In what follows I will attempt to show that whatever the authors' ideas about equality might be, when it comes to the problem of race and IQ they have gotten things backward. The categorization of people by such external physical attributes as skin color, hair form, and nose shape leads to discrimination in education, housing, medical care, and hiring, all of which effect performance on IQ tests. This effect can be reversed through the fostering of equality of opportunity in these areas of social life precisely because race is real not in a biological sense but only as a sociocultural phenomenon. As far as the question of racism is concerned, I leave it up to the reader to decide where the individuals treated in this book really stand in this respect.

CHAPTER TWO

GENETICS AND EVOLUTION

Research on racial differences is frequently based on flawed and/or problematic assumptions about evolutionary theory and population genetics. Therefore readers unfamiliar with the fundamentals of evolutionary and genetic theory will need some guidance in order to fully understand the analyses and criticisms that make up the bulk of this book.

DARWINIAN EVOLUTION

Human variation in all its forms, including so-called racial differences, is, like all biological phenomena, based on evolution, a complex process involving not one but a series of multiple factors.

Charles Darwin's theory of evolution as it has been modified since the publication of *The Origin of Species* provides us with an explanation for the development and diversification of life forms both within and across all species, including our own. Darwin's first major assumption was that all life is related. The number of species occupying the earth has increased through time as a result of continual branching and development from ancestral forms. Evidence for this assumption has been gathered by paleontologists. Their investigations of the fossil record have provided considerable information about the historical development of life on earth. The record shows a progressive development and branching of organic forms.

Given this, the theory proceeds to account for the process of diversification. Darwin (and Alfred Russel Wallace, the codeveloper of the theory) suggested that the variation that exists in nature within *and* between species is *exploited* by variations in the natural environment. Simply stated: Where competition exists for such things as space and/or food, those organisms most fit to survive and *reproduce* in a particular environment survive and reproduce in greater numbers than less well-adapted forms. If the competitive process continues unabated through time, the less well adapted forms are reduced to insignificant numbers or entirely eliminated from the population. Darwin called this process "natural selection."

The theory of evolution includes two sets of balanced biological phenomena. One set provides relative stability in plant and animal species from generation to generation. Another set contributes the source of variation. The first I call "mechanisms of continuity" and the latter "mechanisms of variation." While both sets are necessary for evolution, it is one of the paradoxes of biology that while the mechanisms of continuity reflect the stability of biological systems, the mechanisms of variation are nothing more than mistakes or errors in the process of replication (i.e., the production of the next generation through either asexual or sexual reproduction). These mistakes occur in the germ plasm and are known as *mutations*. Without them there could be no evolution, and accident is, therefore, at the core of the evolutionary process. What this means is that change does not occur in response to need. Evolution is not teleological. It is not based on some grand design and nature does not provide species with the inherent ability to adapt to environmental variation.

At first it was thought that evolutionary changes were exclusively gradual and small. They were supposed to develop *only* if some of the variation already present within a population had a certain value as far as adaptation to new or existing conditions was concerned. Once an adaptive trend has been established, however—that is, once a group of organisms has begun to shift toward a particular type of adaptive change—this change was supposed to continue in the same relative direction so long as the environmental demands remained relatively constant and variations in the direction of adaptation continued to occur in the germ plasm. If, for example, the development of an opposable thumb provided a group of tree-living animals with a distinct advantage for holding on, rather than falling off, a limb, any mutations that produced a better grasping hand would have a selective advantage and a trend would be established. Thus a series of accidental or random events could lead to a nonrandom series of changes.

The above ideas are, in fact, correct but not complete. It has lately been suggested that accident plays another, sometimes bigger, role in evolution, as well. Two paleontologists, Stephen Jay Gould and Niles Eldredge (1993), have suggested that species remain relatively constant through time. According to them, relative stability of life forms is the general rule. They suggest that the majority, and perhaps all *large* evolutionary changes (for example the extinction of the dinosaurs and the rise of mammals) occur when climatic or geological events (ice ages or major volcanic eruptions, for example) accidentally disrupt the environment in major ways. This disruption may randomly affect the number and distribution of existing life forms. Such events then provide opportunities for new, locally adaptive evolutionary change to occur. This general stability, broken now and then by radical changes in life forms, is known as *punctuated equilibrium.*

Even if punctuated equilibrium turns out to be the correct explanation for major jumps in evolution, the adaptive process also occurs as an accumulation of slow changes in existing forms of life. In these cases the environment selects from the total of existing variations only those forms that are best suited to prevailing conditions. When this occurs the possibilities for change are eventually narrowed. As adjustments to a particular environment continue, the chances of widely divergent variations surviving decreases. These will have increasingly *negative* survival value. Adaptation through time represents a goodness of fit between the animal or plant species and the environment in which they live. Another way of putting this is to say that a group of organisms never invades a new environment fully adapted to it. Adaptation involves interaction between the environment and certain biological characteristics of the organisms in question. Conversely, the better the fit between an organism and its environment (the greater its *specialization* in relation to that environment) the less likely that the adaptive process will change course should environmental conditions change.

A final caution: Selection is a local process. Although whole species do change through time, adaptation *only* occurs on the level of local breeding populations (groups of organisms sharing an environment and breeding with one another). Adaptation becomes species wide only through the interbreeding of populations of the same species, a process known as *gene flow*. Gene flow will be discussed later in this chapter.

EVOLUTION AND GENETICS

The mechanisms of variation and continuity can be fully understood only with the help of genetics, a science that was unknown in Darwin's time. Unfortunately, while the original theory of evolution is quite simple, genetics is a complicated business. This is true partially because the theory of evolution is very general, and genetics, as it applies to real situations, is very specific.

The contribution of genetics to evolutionary theory lies in its detailed accounting for the sources of variation and continuity that occur within all living organisms. Combined with the study of environment and the fossil record, genetics helps us to understand not only what has occurred in evolution, but how and why it happened.

First, let us consider some examples of continuity and variation as they occur in familiar situations. It does not take an expert to predict that mating between purebred dogs will produce offspring very much like, but not exactly like, their parents. It would indeed be surprising if a properly mated Great Dane were to labor mightily and bring forth a Chihuahua, or a fox terrier for that matter. On the other hand, crosses between two different varieties of dog will produce offspring of greater variation, and two mongrels of unknown origin will produce quite an unpredictable array of puppies. In all these cases, however, there is a considerable degree of continuity as well. All dogs produce dogs and not cats, and Great Danes produce dogs that are also Great Danes. Breeds, or subspecies, can be interbred to produce mongrels, but species generally cannot be crossed to produce intermediate varieties. In those few cases in which this rule is broken (tigers can breed with lions and horses with donkeys), the offspring of such unions are infertile. Thus the species is a closed unit while the subspecies is not. Human populations, by the way, as we shall see later in greater detail, are more like mongrels than they are like pure breeds.

The source of much of the observed continuity between parents and offspring is *hereditary;* that is, the reproductive process involves the transfer of invariant *genetic* material from parent to offspring. Each breed of dog, however, has thousands of units of this material, which controls such traits as size, shape, color, length of hair, etc. In some cases a single unit, or *gene,* controls a trait; in other (perhaps most) cases, many genes work together to produce a trait. Within any species some traits are held in common by all members of the species, while others are specific to specific breeds. The distinctive attributes of dogs are the result of genetic units distributed throughout the species. The distinctiveness of Great Danes is the result of units distributed within a single breed. These units are generally passed on unchanged, but in sexually reproducing species an offspring receives only one half of its genes from each parent. When two purebred dogs of the same breed mate, much of the genetic material passed on to the offspring from each parent is similar, and little variation results. When mongrels mate, the situation is somewhat different. Genes are passed on unchanged to the offspring, but these genes themselves represent a wider variation in traits because some represent one breed, some another, and some still another. These units combine at random in the puppies, and greater variation results. It is important to emphasize that usually the genetic units themselves do not change from one generation to the next. It is only the combination of these units that is different. The observed variation is due to new combinations of invariant genetic units.

But this is not the whole story. A further complication emerges when we examine variation more closely. A good deal of difference can be produced among litter mates by varying nonhereditary conditions, such as the amount and kind of food given to different dogs, or the amount of exercise allowed to each. When we vary these conditions we are changing the *environment.* Any variation that results from this type of manipulation is the result of environmental factors. It is often difficult to sort out which differences among organisms of a single species or subspecies are due to environment and which are due to heredity. One way of separating these two factors is based on the observation that differences due to the environment acquired during the lifetime of the animal are not passed down to the next generation. This is another way of saying that acquired characteristics are not inherited. Breeding experiments can reveal which traits are hereditary. Thus we know that boxer dogs have been deprived of most of their tails and a good part of their ears for as long as boxer fanciers have docked ears and tails, but that boxers are always born with large floppy ears and long tails. Generations of surgery (a form of environmental change forced on the organism) have not altered the heredity of boxers. If an owner wants a boxer to conform to the show standards for the breed, the dog must be altered in a clearly nongenetic way.

Any differences produced by environmental variation fall within definite limits, for while the environment can strongly influence the development of an individual animal, it can never transform it into something that lies beyond the boundaries of its own heredity. Development, then, is determined by a combination of relatively constant factors that are part of the organism's own potential, and by the external conditions that make up the life experience of that organism. Thus every creature is a product of its own particular environmental history and a part of the genetic history of its ancestors recombined in a unique way.

Three examples of the interaction between heredity and environment can serve to clarify what I mean here. Under normal conditions, one of the distinguishing features of Siamese cats is a type of pigmentation in which the animal's body is generally light in color, with progressive darkening towards the tips of the feet, ears, and tail. This pattern comes chiefly in two varieties, known as "seal point" and "blue point." When cats that look Siamese are mated they will always produce kittens of the same general color pattern. The chemical process that underlies the particular pigmentation of these animals is highly sensitive to temperature. Siamese cats are darker at the extremities because it is these areas of the body that are coolest.

A less trivial example of the interaction between environment and heredity is that of diabetes in humans. This disease is related to a breakdown in the functioning of the pancreas, an organ that, among other things, operates in carbohydrate metabolism. Few people are born diabetic, except for cases of juvenile diabetes, and yet diabetes is recognized as hereditary. What is actually inherited, at least in one form of diabetes, is a weakness in pancreatic structure that may break down, particularly if an affected individual overindulges a taste for carbohydrates. The condition is a result of environmental stress on a potentially diseased organ.

Furthermore, since the discovery of insulin, the hormone involved in carbohydrate metabolism, a patient's internal environment can be altered artificially through insulin injections. These injections restore correct body processes and suppress the effects of the disease.

Some animals, for example the chameleon, have a peculiar hereditary mechanism that is keyed to environmental stimulation. These animals are much more variable than Siamese cats, and chameleon watchers often catch them in the act of changing their color markedly from one moment to the next. For a long time it was believed that these pigment changes were a protective response to background color and that the animal had a built-in device for instant camouflage. Actually, these changes are due to other types of environmental stimulation, such as light intensity and temperature changes, as well as excitement caused by such external conditions as fright. These environmental variables trigger a hormonal response in the organism, and a color change results. Exactly what these reactions have to do with survival is still a debated question, but the fact remains that a chameleon's color at any particular time is a function of inherent body chemistry *plus* internal and external environmental conditions.

Siamese cats and chameleons each in their own way demonstrate that the interaction pattern between genetic and environmental variables may be highly plastic. This is not always the case, however, and, in fact, organisms can be irreversibly shaped by environmental factors. For example, an animal that has been stunted during its normal period of growth will never recover its hereditary potential, and "normal" growth is itself just as much the result of early external influences: in this case, those which maximize the genetic growth potential of the organism. The result of early environment-heredity interaction can set the pattern for the entire life span of the individual.

GENOTYPE AND PHENOTYPE

In order to understand the processes we have been discussing so far, the genetic background must be separated from environmental effects. This can be done experimentally by raising organisms in the same environment (i.e., holding environment constant) and noting genetic variation, or by producing pure strain organisms with practically identical heredity, differentially manipulating the environment of samples of these organisms and noting its effect on variation. Technically the genetic background of an organism is called the *genotype.* But it is important to remember that the genotype only represents the hereditary *potential* of an organism; some genes are activated at specific times only and others, depending on environmental conditions, are never activated. Heredity acting in combination with a particular environment's effect on the genotype produces the *phenotype,* or product of such interaction. Another way of expressing this is to say that the phenotype is the result of a particular heredity acting within a particular environmental background. Any variation we observe among the members of a related group of organisms living under natural conditions must be phenotypic variation, because it will be the result of different environmental pressures and different genetic histories acting together. Phenotypic variation in a population is the sum of genotypic variation plus that part of environmental variation that affects the phenotype.

From this one might surmise that a high degree of phenotypic variation in any given population will occur as the result of high genetic variation, high environmental variation, or a combination of both. Indeed, this is often the case, but it must be noted that the genetic background itself can determine how much and what kind of environmental variation the phenotype can absorb. Some species are highly susceptible to environmental differences; others can remain stable under a wide range of conditions. Stability can result from two different genetic processes. In some species even small environmental variations are not well tolerated. If a population of such a species is exposed to changed conditions, it will die out. Many microorganisms (bacteria, for example) are so sensitive to conditions such as level of acidity or temperature that they will tolerate only minute differences in these variables. By contrast, the genotypic background of other species may be geared to absorb a wide range of environmental variation without developing changes in the phenotype. Humans, relatively speaking, are this type of organism.

What I have been talking about is natural selection, for natural selection represents the effect of environment on the phenotypes of specific sets of organisms, a population. Within the range of variation, those organisms that are phenotypically best suited to the environment will have a selective advantage over the others. This does not mean that every single member of a population that has a selective advantage will survive, or that every member with a selective disadvantage will perish. No biologist would care to predict which individual organisms in a population will live to reproduce, since it is impossible to predict all the events that will occur in the life of an organism. Chance events (accidents) happen even to the best-adapted creatures. A few of the least well adapted may be effectively protected by accident, as well. What the biologist attempts to predict is the success value of a particular phenotype occurring in a group of organisms in which variant phenotypes are present. This means that if two or more types exist in the same environment, it may be possible to identify which type will survive in greater numbers than competing types.

Thus, while all organisms, as individuals, are the products of evolutionary development, evolutionary theory makes predictions about populations and not about individuals. Statements made about natural selection are statistical predictions because they involve probability and are never absolute. They can never give an investigator the assurance that a particular individual will survive. Such predictions represent a percentage figure based on the expected selection value for a specific trait or phenotype. Of course there are situations in which a trait produces 100 percent fatality at birth. In these special cases it can be said with certainty that any organism born with such a trait will not live to reproduce. Its reproductive potential is said to be zero. This fact does not defy the rule that one cannot predict which organisms will live to reproduce, however. It should be obvious from the above that selective advantage is always a comparative figure. There is no such thing as an absolute selective advantage because selection is measured by comparing the reproductive potential of one phenotype against another phenotype or phenotypes always in the context of a specific environment.

SPECIATION AND GENETICS

One of the great questions answered by Darwinian theory is that of how speciation occurs—how one species transforms into another. We know that species cannot interbreed and produce fertile offspring, so some other mechanism or set of mechanisms must be involved if Darwin's principle that all life forms are related is true. Let us examine this question by assuming that a sexually reproducing relatively homogeneous species lives within a small restricted environmental zone. Let us further assume that this species constitutes a single breeding population. This means that any sexually mature member of the population has as good a chance as any other sexually mature member to mate with any other mature member of that population of the opposite sex. Such a mating pattern assures a wide random distribution of genes in the population at large. Random mating of this type is known as *panmixis.* When it occurs, a population is said to share a common *gene pool* (the sum of genetic variation found within a breeding population). Any variation that exists in such a population will be distributed fairly evenly within the confines of the total group. Now if this is a particularly successful species and, therefore, it spreads out geographically, it is likely that subpopulations will develop as units of the larger group. If the distance between these groups widens, they will eventually constitute separate breeding populations. This is largely a mechanical situation. Animals that live in proximity are more likely to breed than animals that live far apart. If any barriers develop between these populations, they will become at least partially isolated. In such situations new genetic variations will be unequally distributed in the species at large. That is, each subgroup will begin to develop its own gene pool different in some respects from all the other gene pools. If the geographic space in which the species is distributed is uneven—that is, if there are environmental variations to which the species is sensitive—then different selection pressures will further differentiate the gene pools of the subgroups. As long as some interbreeding continues to occur between these subgroups, no new species will be created. Each individual subgroup will constitute a separate breed or strain (or race) of the species. If, however, for some reason some of the populations become totally isolated, they will continue to change to the point where genetic differences will be great enough to produce new species. The genetic distance between such "daughter" populations will now be so great that they will not be able to interbreed and produce fertile offspring. Under natural conditions subpopulations tend to become separated through such centrifugal processes as differential genetic variation, differential selection pressures, and semi-isolation. They are also frequently drawn together by the centripetal process of gene flow caused by even occasional interbreeding among adjacent populations. Speciation occurs when the centripetal forces are interrupted.

Now let us imagine our example species to be widely distributed in space so that several subpopulations are scattered over a large area. Assume that we are interested in gene flow and differentiation as they occur in a lineal direction from one end of the geographic space occupied by the species to the other. In such a situation it would make sense to assume that gene flow occurs as genetic transfers among adjacent populations. If we label our adjacent groups A, B, C, D, E, . . . n, then it should be obvious that some interbreeding will occur between groups A and B, between B and C, between C and D, etc. Genes from A get to population n not directly through interbreeding, but via a chain of breeding among adjacent populations. If each group is a partial genetic isolate—that is, if most of its breeding takes place within a group—one would expect to find a gradual diminution of genes from A as these pass through B, C, D, E, to n. A particular genetic trait with a high frequency in A would have a lower frequency in B, still lower in C, and so on. If, in addition, environmental pressures were different for each population, there might well be a further diminution of the trait as one moved from an area of high selective advantage to an area of low or negative selective advantage. Such a gradient is known as a *cline.* Clines are quite common. For example, gene frequencies (the percentage of variants found in separate populations) of the various human blood groups in the ABO family vary clinally across the Eurasian landmass. Blood groups A and B are more common in Asia than in Europe. The same is true for the Rh system, with Rh negative found more frequently in East Asia and diminishing in frequency as one moves westward. Another cline, this one for earwax in East Asians, is distributed in the north-south direction, with a higher percentage of yellow sticky wax closer to the equator and more dry and flaky wax as one moves northward.

Clinal distributions tell us something about populations of a species, for while clines refer to the distribution of a single trait in a species they often reflect a scale of difference between populations. When the clinal difference between A and n becomes very great, it may well be impossible for individuals from A to produce fertile offspring when mated to individuals from n. This would be true if many other gene differences had developed along the clinal distribution between A and n. It is important to note that no example of this degree of clinal differentiation has occurred among human populations. All human populations can breed with any other human population.

Aside from the human case, however, this presents a problem. When clinal differentiation has reached a high degree, as it does in some nonhuman species, are populations A and n members of different species, or do they represent variant strains of the same species? The answer depends upon which evidence one might wish to employ in the genetic definition of species. In the large sense A and n are members of a common gene pool even if each individual unit of the cline pool constitutes a subpool. In the case under discussion, genes can only be exchanged between A and n via intermediate populations genetically more similar to one or to the other. If, in the type of clinal distribution discussed here, one population, let us say C, is wiped out, then the gene flow between A and n will become impossible. As a result the genetic difference between A and n will widen at a greatly accelerated rate. Furthermore, with the disappearance of population C, no connection between A and n will exist to link them naturally as former members of a common gene pool. Populations A and n are now separate species.

MENDELIAN GENETICS

Since the earliest domestication of plants and animals, humans have used certain genetic principles without fully understanding them. Species were improved artificially through selective breeding in which animals or plants with desired traits were bred to one another.

To understand how selective breeding actually works, one must understand how and in what frequency specific characteristics pass from parent to offspring. Traits do not always appear to pass in orderly fashion. Two parent animals, each with the same trait, might transfer it to all their progeny while two other parents, again with the same trait, might transfer it with a frequency of only 50 percent. By contrast, an entire generation of siblings might differ considerably from both parents. Until the reasons for these variations were understood it was difficult to predict what the outcome of any specific cross might be.

A set of simple but elegantly controlled experiments conducted by the Abbe Mendel in the 1860s solved this basic problem and later opened up the entire field of genetics. Mendel's experiments were carried out on the common pea plant. In his search for order in the transmission of genetic material, Mendel chose plants displaying seven classes of easily observable traits, each with two variants. Among these were: height (tall vs. short); seed color (green vs. yellow); and seed texture (smooth vs. wrinkled). Mendel was careful to choose plants of pure strain, which, when crossed with like plants, bred true consistently. Thus, yellow plants produced 100 percent yellow offspring; short plants produced 100 percent short plants, etc. Plants with various combinations of these three classes were then crossed (tall, green, and wrinkled with short, yellow, and smooth, for example) to produce a filial or F_1 generation. Members of this generation were again crossed to produce still another filial or F_2 generation. After the completion of each cross, the frequencies of the resulting traits were carefully recorded and analyzed statistically.

SEGREGATION

Mendel's first concern was with traits of a single class. In these experiments he found that one trait in each class would fail to appear in the F_1 generation. A yellow crossed with a green, for example, yielded an F_1 generation with 100 percent yellow-seeded plants. Furthermore, when F_1 plants were crossed with one another, the green color would reappear in 25 percent of the F_2 offspring; the other 75 percent would, of course, be yellow. It appeared that yellow was able to mask the green, at least in the F_1 generation. Mendel accounted for this fact by suggesting that some traits were *dominant* over others. The reappearance of the green component had "segregated out." Traits that could be masked in this way were said to be *recessive*. Mendel also found that in no case did a cross between green and yellow, tall and short, or smooth and wrinkled produce an intermediate form. Traits might be masked, but they never blended.

The reason for the specific frequency of the recessive trait in the F_2 generation can easily be accounted for by considering the simple permutations and combinations possible when two pure types are crossed. Each parent plant will contribute one genetic unit, or gene, for the trait in question. Just as obviously, this must mean that the offspring will contain two units (one from each parent) for the trait. Different units of the same trait class—tall vs. short in the class of height, for example—are called *alleles.* In this example they are alleles (variants) of the height gene. Every plant contains two units for each trait, only one of which will be donated to a particular offspring. These alleles are donated at random. This means that each allele from each parent has as good a chance as any other allele to turn up in the offspring. Now if we cross two pure strain plants of opposite type—a yellow with a green, for example, we can represent the cross as YY (yellow parent) X gg (green parent). All of the offspring of this mating will be the mixed type Yg since all possible permutations and combinations yield identical products (Yg, Yg, Yg, Yg). All the offspring from this mating will be yellow and, therefore, the allele for yellow is said to be *dominant* over the allele for green that, by the same logic, is called a *recessive* allele. Now if we cross any two of the F_1 yellow plants we should expect the following results. YY, Yg, Yg, gg. Twenty-five percent of the offspring will be gg (recessive) and therefore green. In order to simplify the notation system in genetics, the letter symbolizing the dominant allele is used to symbolize both dominant and recessive alleles, the dominant form in uppercase and the recessive in lowercase letters. Thus our F_2 cross would be written as follows: YY, Yy, Yy, yy. Here the yy combination stands for the 25 percent green seeded, recessive offspring.

When the two alleles of a particular gene are the same (YY or yy) the organism is said to be *homozygous* for that trait. When the two alleles are different, as in Yy, the organism is said to be *heterozygous* for that trait. Heterozygotes are also referred to as *hybrids.*

Mendel's experiments revealed another genetic regularity that he named the law of independent assortment. This law is concerned with the relationship among classes of traits (height or color, for example) rather than between traits of a single class. When Mendel crossed F_1 hybrids, he found that these various classes combine at random in the F_2 generation. Thus, if you begin with only tall yellow and short green plants, you will eventually end up with tall green and short yellow plants as well.

CHROMOSOMES

When one realizes that many thousands of traits contribute to the makeup of an organism, it should not be difficult to realize also the amount of potential variation present when genetic traits follow the law of independent assortment. There is, however, a restriction on the law that limits variation to a considerable degree. If all genes actually existed as independent units, like separate beads in a jar, variation would be totally dependent on the laws of chance. Genes are, however, actually more like strings of beads: They occur in sets, strung out on long molecules called *chromosomes*. Every living cell (plant and animal) contains several of these chromosomes, and their number is determined by the particular species in which they occur. Humans have twenty-three pairs of chromosomes, totaling forty-six. Eggs and sperms, of course, have only one set of each pair, or twenty-three chromosomes in total. Mendel was extremely lucky. Every one of the traits he used in his experiments occurred on a different chromosome, and therefore acted independently according to the law of independent assortment.

CODOMINANCE

But Mendel's luck was even greater than this. For not only can the law of independent assortment be violated by chromosome linkage, but the fact that genes are independent units that do not blend to produce intermediate forms can *appear* to be violated as well. Everyone knows, I think, that there are certain crosses in which it is possible to combine what appear to be variant forms of a trait class. In certain species of plants, for example, it is possible to cross white- and red-flowered plants and produce an F_1 generation of pink flowers. It is easy to prove, however, that the alleles in question have not mixed together to produce the pink result. All one need do is cross F_1 plants together. The result in the F_2 generation will be 25 percent red plants, 50 percent pink plants, and 25 percent white plants. The red and white alleles acting together do produce pink, but they do not *mix* together and lose their independent qualities. When alleles influence one another in this way we call them *codominant.*

POLYGENES

There is another case of blending inheritance, however, that is more complicated. This kind of blending occurs when instead of one gene (with its two or more alleles) for a trait there are several genes at different places *(loci)* on a chromosome or even on different chromosomes. Such traits, and there are many of them, are said to be *polygenic.*

Let us now imagine a case in which height is dependent upon three gene loci. Assuming that there are dominant and recessive alleles for each loci, a number of possible genetic combinations will produce different heights in an F_1 generation that is the result of a hybrid cross. If the tall genes are dominant over the short genes, then the tallest individuals will have at least one dominant allele at each locus. On the other hand, the shortest individuals will be homozygous recessive at all three loci. Between these extremes there would be a distribution or range of heights, bounded only by the tallest and shortest possible combination of alleles. The situation would be even more complicated with codominant alleles at each locus. In such a case the tallest individual would have to be homozygous tall at each locus (six tall alleles) and each short gene present would have some effect on total height. The existence of polygenetic effects is demonstrated through breeding experiments that show a range of variation within a trait class.

GENE MODIFICATION

A further complication of the polygenetic effect results from the fact that some genes may not only have an additive effect but can also modify or suppress the action of another gene at a different locus. This effect is known as *epistasis*. A gene that can suppress the effect of another gene at another locus is said to be epistatic to that gene. (The suppression of a recessive allele by a dominant allele at the same locus is a special case and is not called epistasis.) The suppressed gene is said to be *hypostatic* to the suppresser gene. A clear example of this effect is found in Siamese cats. The seal-point Siamese has a normal gene for dark color at one locus and is homozygous for a recessive allele at another. It is these recessive alleles that modify the cat's color. The blue-point Siamese has the same suppresser alleles, plus still another set of alleles at another locus that modifies the color gene and dilutes its effect. This allele acts as a dilutant to normal color.

Albinism in animals and humans is usually due to the epistatic effect of a recessive suppresser gene acting on the normal color locus. Thus the genes for eye, hair, and skin pigment are all hypostatic to the gene for albinism.

SEX LINKAGE

Most sexually reproducing species are *dimorphic,* that is, of two types, which are recognized as male and female. The determination of sex is a genetic process based on chromosome difference. The chromosomes responsible for sexual differentiation are the only ones that are not homologous. One of the pair is shorter than the other, and for this reason they can be readily identified microscopically. In humans, the short chromosome occurs only in males and is known as the Y chromosome. In humans, female sexual chromosomes are both of normal length and are known as X chromosomes. When sperms form in the male they are either X or Y. Since females can only produce X chromosomes, their eggs will always be X, and hence a woman cannot determine the sex of her offspring. Since the Y chromosome is shorter than the X chromosome, it is possible for traits located on the nonhomologous part of the X to express themselves in a single dose. Such traits are said to be *sex-linked.* Color blindness in humans is a good example of a recessive sex-linked trait. A man can be colorblind with only one allele for the trait, but a woman must be homozygous recessive to be colorblind. Since the allele that causes color blindness is rare, there are bound to be many more colorblind men than women. In fact colorblind women are very rare. Sex-linked genes can, however, be dominant as well as recessive. When this is the case, twice the number of women than men is affected. In this case a man has only one chance to get the allele while women have two chances.

PLEIOTROPY AND PENETRANCE

Genes are chemical substances, part of a long molecule known as DNA. The structure of DNA will not be discussed here, but it is important to note two chemical facts that have to do with how genes produce traits in organisms.

It must be stressed that the genes that make up DNA produce proteins and not phenotypic traits, at least not in the direct sense. A long series of chemical reactions takes place in the organism when a gene is activated. *Only the end product* of these reactions gives rise to a trait. Genes interact with one another and with the organism as they come to be expressed in the final product. In this process a gene may affect not merely one but a series of traits, resulting in what can be called a *trait complex.* This is the case because a single protein activated by a gene may lie at the base of various chemical pathways that themselves produce a host of phenotypic responses. The multiple effect of a gene is called *pleiotropy.* Most, if not all, genes are pleiotropic.

I have already noted that the phenotype is the result of complex interactions between alleles, genes at different loci, and the environment. Depending upon conditions that may be either external or internal to the organism, a particular gene may be expressed along a continuum. This can range from a total lack of expression to the expression of maximum effect. The degree of expression of a genetic trait is referred to as *penetrance.* If for some reason a normally harmful gene has low penetrance then its effect on the survival of a population will be less marked than if it had high penetrance. Again it is necessary to emphasize that genes act in combination with environmental pressures and not as isolated determinants. The genetic aberration that produces diabetes, for example, might have a penetrance near zero in a population whose diet was extremely low in carbohydrates. The presence or absence of the gene for diabetes in such a population would have little effect on fertility or mortality and, therefore, could not be said to have a low coefficient of selection. If the same population changed its eating habits, however, the combined effects of the gene plus the new environment (eating patterns are part of the environment) might reveal a rather high negative selection value for the gene.

The combined effects of penetrance and pleiotropy help to make selective value a flexible proposition. Environmental factors might well suppress or modify one action of a gene and have little effect on another. Any statement about the selective value of a gene must, therefore, always be given in terms of a specific context.

POPULATION GENETICS AND EVOLUTION

The evolutionary process cannot be understood without analyzing the effects of genetics and environment as they interact in breeding populations.

The theory of evolution presupposes that beneficial changes in genetic structure are preserved in living forms and passed on through the reproductive process to the next generation. This means that within any species there must be a certain rigidity that operates to preserve evolutionary gain. If further adaptation is to occur there must also be a certain amount of flexibility present to provide the material for change. To speak of the rigidity or flexibility of a species, however, we must refer to a group within which there is a sharing of genes through the mechanism of sexual reproduction. Most species cannot be considered in this way, for they themselves are made up of separate spatially distributed breeding populations. The boundaries between the populations can be considered to be the boundaries between distinct gene pools. This does not mean that breeding is totally confined within such groups, for some gene flow must occur if the integrity of the species is to be maintained. It is convenient and necessary, however, to construct a simple model of a breeding population that does not conform totally to any actual situation. This allows us to project more realistic and complicated problems against a set of general abstract principles. Such an abstract model of a population can be manipulated mathematically by introducing variables at first one at a time and then in different combinations to yield sets of expected results. These can then be tested in the laboratory or through observation of natural populations within which a known set of events is occurring. Those events are of the following kind.

MUTATION

As we have seen, mutations are chemical changes in genes that may have a positive or negative effect on a trait in relation to the environment in which the population is imbedded. Under most conditions mutations occur at a low constant rate and are reversible. That is, if a gene mutates from A to a it will also mutate back from a to A. The rates of back mutation, however, need not be the same as the change from A to a. The mathematical model of population genetics tells us that if a mutation is new (a rare event) the new allele will build up in the population until such time as a sufficient number of genes exist that mutate back to A, establishing a new equilibrium between mutations and back mutations. Thus even in populations with constant mutation rates, the frequency of genes will eventually come to equilibrium as long as mutations occur in both directions. From the point of view of evolution, the fact that equilibriums are established between opposing mutations means that under stable conditions where selective factors are *inoperant,* variant alleles are maintained at a constant rate within the gene pool of the population.

SELECTION

The effect of selection on the population model is more complicated than the effect of mutation. This is the case because selection acts on the phenotype and not directly on the genes. In a two-allele model, one must consider three possible genotypes and the degree of penetrance of each. If a gene is completely recessive, selective forces can act on it only when it occurs in the homozygous state. If genes are codominant, the effects of natural selection will be different for the homozygous and heterozygous state. Dominant genes, on the other hand, will be affected equally in both states. In addition to these differences, there are important cases in which a heterozygous phenotype will have selective advantage over either type of heterozygote. This is known as *overdominance.* Overdominance can lead to a situation known as polymorphism in which several alleles of the same gene are maintained in a population at high frequency.

In most cases selection favors one allele at a locus at the expense of others. If selection has been operating on a locus for some time, the geneticist expects to find the most fit allele in high frequency and the less fit one in low frequency. There are situations, however, in which two or more alleles of the same gene occur in high frequency. At first it might seem as if selection were not operating in these cases. This phenomenon may be due, however, to a condition of selection in which the heterozygote is favored over either homozygote. Such a situation leads not only to a high frequency of heterozygotes, but also to some sort of balance between the dominant and recessive alleles, since both must be present in high enough frequency to assure the presence of the Aa combination.

An interesting example of this phenomenon is found in human populations. The chemical structure of normal hemoglobin is controlled by the dominant gene HbA. A mutant gene, Hbs, also occurs. This allele in the homozygous state produces a disease known as sickle-cell anemia, which is generally fatal in the early years of adulthood. The allele Hbs has low fitness, since homozygotes will have the disease and, therefore, lower fertility than normal individuals. Heterozygotes for this condition normally also have a somewhat lower fitness since their red blood cells are abnormal and transport oxygen less well than completely normal cells. Their fitness, however, is closer to the homozygous dominant than it is to fully affected individuals. Some years ago it was noticed that the allele Hbs occurs in high frequencies in certain populations in West Africa. This finding is surprising because sickle-cell anemia is just as severe in Africa as it is elsewhere. Later it was discovered that heterozygous individuals (HbA, Hbs) were resistant to a severe form of malaria (falciparum malaria) found in the same region as the high (Hbs) frequency. This resistance appears to be related to the effects of the sickle cell gene on the red blood cells since the malarial parasite spends a good part of its life cycle in these cells. In regions where this type of malaria is common, heterozygotes have an advantage over both types of homozygotes. The homozygote recessive (Hbs, Hbs) succumbs to sickle-cell disease and the homozygote dominant (HbA, HbA) succumbs to malaria. The heterozygote (HbA, Hbs) is resistant to malaria and does not display severe symptoms of sickle-cell anemia. The heterozygote advantage of HbA, Hbs probably developed after the introduction of agriculture into West Africa, when a pursuant change from a closed forest environment to open sunny areas occurred. This change, dependent upon culture, favored a species of mosquito known to carry the falciparum parasite and that breeds only in sunlit stagnant pools of water. It is also known that the Hbs gene originated in East Africa among Bantu-speaking peoples and was carried to the West by migrating populations. Populations with a longer West African lineage have a lower frequency of the s allele than those whose ancestors moved from East Africa.

GENETIC DRIFT

There are three other phenomena that have a direct effect on the frequencies or distributions of genes in a population. These are *genetic drift, interbreeding,* and *inbreeding.* Genetic drift is merely the result of sampling error, occurring, for example, when a small segment of a population migrates away from a parent stock. If the migrating population is a true random sample of the original population, and if it is large enough, one would expect it to have the same gene frequencies as the parent population. If the migrating population is small, however, it is unlikely to be a faithful representation of the parent population. When this is the case the new population may differ significantly from its "parent." This is known as the *founder effect.* Small populations are also subject to genetic drift through time. If a catastrophic event such as a tropical storm were to wreak havoc on the population of a small island and kill a significant number of its people, the survivors would likely have a gene pool different from the prestorm population. Such a change is known in the jargon of genetics as the *bottleneck effect.*

Small populations in general are subject to nonrandom accidental changes in gene frequency that may have little or nothing to do with either selection or mutation. In addition, social regulations that affect the potential reproductive capacities of individuals may interfere with their actual biological fitness. Thus, a wealthy man in a polygamous society may have many wives and children while a poor man in the same society may produce few offspring. Celibacy imposed on a segment of a group through, for example, religious beliefs may reduce the fertility of potentially fit individuals to zero. The fact that human groups were small throughout most of history, coupled with the fact that social rules differ markedly from group to group, has certainly served to widen the genetic gaps between populations. This divergence is, however, subject to the reverse process through gene flow among potentially divergent populations. The human species has always been extremely mobile, adaptable through culture to a variety of environmental conditions. It is unlikely that many true population isolates have existed except in unusual circumstances, and then for relatively short periods. It must be remembered too, that selection is usually a very slow process and our species is, in evolutionary terms, a young one.

INTERBREEDING

Interbreeding among populations that differ markedly in genetic constitution presents yet another case of genetic modification. A cross between two such groups creates a third, intermediate group that presents a different face to the process of natural selection. Interbreeding brings about changes in gene frequency and type just as do mutation and natural selection but, of course, interbreeding is a much faster means of producing change than the other evolutionary processes.

SEXUAL SELECTION

While Darwin is known as the father of evolutionary theory and the discoverer of its major mechanism, natural selection, he was aware that other types of selection could also occur in sexually reproducing species. Mate choice on the basis of observable genetically programmed traits could radically change the outward appearance of both male and female members of a population. This type of change is known as *sexual selection.* Sexual selection occurs when mates are chosen by individuals on the basis of some preferred external, genetically controlled trait. It has been used to explain the spectacular plumage found among males of several species of birds as well the extravagant antlers seen in some deer and elk species. Sexual selection can only change the outward appearance of organisms, their "package." It has little or no effect on the internal genetic structure of an individual, their "contents."

BEHAVIORAL GENETICS

If genetics controls physical traits can it also control behavioral traits? While extreme caution must be taken in dealing with this question, the answer is "yes." Professor Jerry Hirsch, working at the University of Illinois, is one of the pioneers in this field. Working with the once-favorite laboratory animal of geneticists, the fruit fly *Drosophila,* professor Hirsch demonstrated that the normal behavior of these organisms to fly upward against the pull of gravity when placed in a confined space could be reversed by artificial selection. Hirsch constructed a maze that consisted of a series of alternate choice points. At each point an insect could fly either down or up. When placed in the maze, most of the wild flies drifted toward the top of the maze through time, as expected. Some of them, however, were found to stay at various levels below the highest point. These insects were then collected, and their offspring run again. After each run, files that exhibited a downward choice were preserved and bred. After several generations, Hirsch's efforts had resulted in a strain of genetically controlled downward-flying organisms. What he had done was exploit a naturally occurring variation within the species (Ricker and Hirsch, 1988).

After the pure strain of down-flying insects was established Hirsch was able to cross them back to the wild strain to determine the genetic structure of the trait. As would be expected, it followed the normal Mendelian rules of inheritance.

Several years ago Professor R. C. Tryon of the University of California at Berkeley tested the notion that laboratory rats could be bred for their ability to run mazes. He constructed a rather complex maze and selected two groups of rats: those that performed well in the maze and those that performed poorly. "Intelligent" rats, that is, those that did well, were inbred and eventually a strain of these animals (known as the S_1 strain) was developed. A strain of poor performers (S_3 rats) was also isolated through artificial selection (Tryon, 1940). For several years it appeared as if Tryon had actually selected out strains of smart and stupid rats (intelligent and stupid relative to some standard of average rat intelligence). In the 1960s, however, a group of investigators at Berkeley showed that the S_1 animals were good at only one particular kind of maze based on spatial cues. In fact the S_3 rats did better than the S_1s in certain other maze experiments. The lesson here, of course, is that it is necessary to be very specific in defining what the selected trait actually is. When working with physical characteristics this is relatively easy. When it comes to behavioral traits an experimenter faces a more difficult problem.

When they first began their experiments on the behavioral genetics of dog breeds, the psychologists J. P. Scott and J. L. Fuller thought that aggression and timidity could be defined along the same dimension. That is, they assumed that dogs would be either aggressive or timid but not both. They soon discovered that aggression and timidity were two separate traits. Thus a dog that attacked, moving toward a stimulus while barking and growling was an aggressive-nontimid dog. A dog that cowered in a corner while snapping at a stimulus was an aggressive-timid dog. A dog that cowered in the corner and whimpered was a nonaggressive-timid dog. This example shows that behavioral genetic experiments demand very careful sets of operational definitions in order to determine the relationship between a defined type of behavior and its assumed genetic substrate. We can generalize from this finding and say that if proper definitions are important for such *relatively* simple traits as aggressiveness then they are even more so for such complex traits as intelligence (J. P. Scott and J. L. Fuller 1951).

CHAPTER THREE

RACE: A FLAWED CATEGORY

In biological classification there are two and only two relatively unambiguous categories. These are the species and the individual. Both are virtually closed units. The individual is a single organism. The species is a unit that can breed with itself but not with another like unit and in so breeding produce fertile offspring. A third unit, the *breeding population,* is less definite than either the individual organism or the species because gene flow among such units can occur. Breeding populations are subunits of a species in which, for various reasons such as geographic isolation or, in the case of humans, marriage rules that forbid mating with certain categories of individuals, all or almost all breeding takes place. The boundaries of breeding populations are at least theoretically permeable. Thus they are not closed units like species or individuals. In spite of this difficulty breeding populations are very important units in evolutionary biology. They are the real locus of evolution. Yes, individual organisms *are* selected by the environment, but the dynamics of the evolutionary process take place among the organisms that constitute a breeding population. Individuals do not have gene frequencies; only populations do. Gene frequencies are statistics that reflect the percentage of one kind of allele of a gene in reference to all other alleles of the same gene. One example would be the frequency of the allele for blood group A in a population in relation to the frequencies of the alleles for blood group B. (The blood group AB contains both A and B alleles.)

The most important fact about the concept of race, whether it is used in biology or common discourse, is the porous nature of the term. For example, the authoritative *Oxford English Dictionary* offers the following (I omit obsolete definitions):

I. A group of persons, animals, or plants, connected by common descent or origin.
  1. The offspring or posterity of a person; a set of children or descendants.
    b. Breeding, the production of offspring.
  2. A limited group; of persons descended from a common ancestor; a house, family, kindred.
    b. A tribe, nation, or people regarded as of common stock.
    c. A group of several tribes or peoples, forming a distinct ethnical stock.
    d. One of the great divisions of mankind, having certain physical peculiarities in common.
  3. A breed or stock of animals; a particular variety of a species.
    b. A stud or herd (of horses).
    c. A genus, species, kind of animals.
  4. A genus, species, or variety of plants.
  5. One of the great divisions of living creatures:
    b. Mankind
    c. A class or kind of being other than men or animals.
    d. One of the chief classes of animals (as beasts, birds, fishes, insects, etc.)
  6. Without article:
    b. Denoting the stock, family, class, etc. to which a person, animal or plant belongs. . . .
    c. The fact or condition of belonging to a particular people or ethnical stock; the qualities, etc. resulting from this.
  7. Natural or inherited disposition
II. A group or class of persons, animals, or things, having some common feature or features.
  8. A set or class of persons.
    b. One of the sexes.
  9. A set, class, or kind of animals, plants, or things.
  10. A particular class of wine, or the characteristic flavor of this. . . .
    b. Of speech, writing, etc.: A peculiar and characteristic style or manner, *esp.* liveliness, sprightliness, piquancy.

Admittedly, the Oxford Dictionary's definition of race is based on current usage; it must, therefore, offer a wide range of sometimes contradictory choices. Many of these, however, can only lead to guaranteed confusion. If, for example, race is a biological concept then how can *ethnic* stock be included in its definition when ethnic stock is a purely *cultural* (not a biological) entity? This is an example of the well-known proverb about adding apples and oranges, which, as we know, can only produce fruit salad!

Scientists concerned with human variation have not done much better than laypersons when attempting to define race unambiguously. Ashley Montagu, a physical anthropologist known to be hostile to the concept of race, dropped the term altogether. He substituted "major group" for "race" and "ethnic group" for "subrace." But, as has just been noted, "ethnic group" is a sociological, not a biological, concept. If it was Montagu's intention to get away from the biological category "race" he was unsuccessful. His "ethnic groups" were just seen as another word for what had been taken as biological races or subraces. If we turn to anthropologists who championed the concept of race we still find major problems. Thus, for example, Carlton Coon in *The Living Races of Man* says the following:

> Not every person in the world can be tapped on the shoulder and told: "You belong to such and such a race" and this fact has made some people think that there are no races at all. Ever since man first spread over the earth, interracial contacts have taken place between the populations of neighboring geographical regions, with constant genetic exchange that has produced racially intermediate, or so-called clinal populations. (Coon 1965, 7)

What Coon would do to remove the ambiguities he mentions is exclude these "clinal" populations from the main subdivisions of our species. The term "cline," however, as Coon admits elsewhere (1965, 212), refers to a gradient in the frequency of a particular trait through a series of populations of the same species. It cannot be applied to populations as a whole. This is because clinal distributions of different traits may vary differently through the same set of populations. Skin color, for example, varies on a north-south gradient, while the ABO blood group system varies from east to west. Equally important, by excluding these "messy" intermediate populations, Coon inadvertently demonstrates why racial groups cannot be defined as closed units. Finding a pure race is like looking for the heart of an onion. You peel away layer after layer until you are left in tears with nothing in your hand. *Yes,* interbreeding is a major factor in confusing the issue of race, and it has been occurring for so long and so constantly that it has created a genetic hodgepodge of the human species. There are just *NO* biologically *pure* races *anywhere.*

Coon's discussion of different races (he believed there are five: white, or Caucasoid; black, or Congoid; yellow, or Mongoloid; Native Australians, or Australoid; red or Amerindian) illustrates another confusion in his thinking. He saw his "own" race (Caucasoid) as the most highly variable of his five divisions. One might expect that his explanation for this would be based on the same kind of gene exchange that produces "clinal" populations. But no, Coon had most of the gene flow going out of Europe, the supposed Caucasoid heartland, *into* Africa and Asia. In fact, in one of his books, he stated that Negroids were the result of a combination between Pygmoid and Caucasoid genes (1965, 123). Although Coon believed Mongoloids to be one of the purest races, with the lowest degree of variation, he attributed a fair dose of Caucasoid genes to them as well.

Coon believed Congoids and Mongoloids, as opposed to Caucasoids, to be highly uniform in physical structure. Perhaps what we see here is a familiar type of culturally biased judgment in which one is prone to make finer distinctions among things (or groups of people) with which or whom one is most familiar. We know, for example that many non-Chinese Americans find it difficult to tell one Chinese from another, but we ignore the fact that many Chinese have the same problem in respect to, say, white Americans. Anti-American propaganda plays put on by Chinese Communists during the Cold War depicted Americans as heavily mustached with pronounced noses. For the Chinese, of whom the majority, but not all, have small noses and less body hair than most Caucasians, these two characteristics submerge any other physical differences that might be used to distinguish individuals or groups.

The fact that the Chinese might have difficulty distinguishing among Caucasians was brought home rather forcefully to my late colleague, Morton Fried of the Department of Anthropology at Columbia University. Professor Fried did his ethnographic fieldwork in China in 1949. Here is his story as he told it many years ago:

> There were very few foreigners in the area, mainly missionaries who would come and go, either visiting the ancient Catholic priest who had almost forgotten his native Italian and lived in his church, hardly ever venturing out, or the younger set that dropped in on the Protestant missionary who had only recently arrived. That missionary, George Cherryhomes, was, at least in physique, the archetypal Texan, well over six feet in height, lean to the point of emaciation, fair of skin and hair. We must have made a strange contrast in the streets of Ch'usien, the only foreigners to be seen there with any regularity, for I am at least half a foot short of Cherryhomes' height, outweighed him by more than 50 pounds at that time, and have a ruddy complexion and brown hair.
>
> Yet, on numerous occasions as I made my way along Chungshan Street or Great Eastern Road, I would feel a tugging at my arm and turning, would look into the face of Mr. Wang, the local postman. Beaming happily, he would bestow a fistful of letters and papers upon me and skitter away. It never failed—the mail thus delivered would be for Cherryhomes.
>
> Finally, I confronted Mr. Wang and asked him why this constant mistake. After all he knew me and he should have known Cherryhomes as well. Wang studied me carefully for several minutes without speaking. Finally he shook his head sadly and said, "Wai-kuo jen tou hsiang I-yang." ("You foreigners all look alike.")

It was my own experience in Africa among a single population in a typically "Negroid" area of the Ivory Coast that the amount of variation in skin color, nose form, and hair color, to mention only three traits, was, to my originally naive eyes, rather startling. Skin color varied from very light brown to dark brown. Nose form varied all the way from broad and flat to aquiline. I even found a fair proportion of redheads and dusty blondes in the population.

While I have already pointed out that the amount of physical variation within any so-called geographic race is considerably greater than most casual observers would perceive, nonetheless it may also be the case that certain *external* phenotypic characteristics (what I have referred to as "packages") do tend to be relatively uniform. That is to say, such features as skin color, hair form, and facial shape may vary less within populations than do other equally important genetic features. While the package may be relatively uniform, the contents are probably not. This is because both natural selection *and* sexual selection can have strong effects on phenotypes.

Where environment varies regularly through space, adaptational clines may develop as populations adjust to these variations. But people can also shape the common *visible* phenotype in their population by actively choosing what physical features they prefer in a mate. Of course they cannot (except in advanced scientific cultures) do anything about the presence or absence of invisible traits. In human societies in particular, fertility can be, and often is, influenced by cultural factors, and sexual selection may well help explain the *relative* consistency of certain external phenotypic characteristics within populations.

But to return to the definition of race. Stanley Garn, a physical anthropologist specializing in human variation, employed the following definition:

> A race in man, as in any living form, is a population, a population of men, women and children, of fathers, mothers, and grandparents. Members of such a breeding population share a common history and a common locale. They have been exposed to common dangers, and they are the product of a common environment. For these reasons, and especially with advancing time, members of a race have a common genetic heritage. (1961 and 1965, 5)

This definition equates "race" with "breeding population" and as such would be acceptable if it did not render the term "race" redundant and confusing. If "race" refers to a breeding population let us call it that and *only* that. Continuing to use the term "race" only perpetuates a faulty concept.

In old-style biology, species and races were classified on the basis of *type specimens* (a single example of a plant or animal that was taken to be typical for the entire group to be classified), stored away in the drawers of museum collections. It was assumed that one need only to study the type specimen to know all there was to know about the group it was supposed to designate. The type specimen allowed for neat and unambiguous categories but had no relationship whatsoever to the realities of the living world. Just as importantly, it was of no use to the study of evolution as a dynamic process.

In the 1950s biologists and anthropologists began to think seriously about the problems caused by the static nature of their classification systems for humans. It was the field of chemical genetics, particularly the study of blood groups and their world distributions, that opened this field of study. The best-known work of this type was done by the biologist W. C. Boyd. His 1950 book on blood group distributions brought about a revolution in thinking about human variation. With the proliferation of field studies of genetics among populations it soon became apparent that race was a flawed concept. Few, if any, populations in the world had exclusive possession of even single alleles. Instead, populations could best be distinguished one from another in terms of varying gene (allele) frequencies. That is to say, most alleles were found to be quantitatively distributed across the entire species with some groups having higher or lower frequencies than others. Instead of neat biological units, each one separate and distinct, the biologist was left with a set of overlapping distributions with different means and a good deal of overlap at the edges.

This new way of looking at human variation was at first not seen as a challenge to the race concept. Rather, it seemed as if it merely needed to be modified with more attention paid to the clustering of certain gene frequencies among related populations. But the investigation of several new traits (blood groups, hemoglobin variants, and some other blood components) began to produce a confusing picture of genetic distribution around the world. With this new data in hand, races could be defined as discrete units only if a major part of the data were neglected. When large numbers of traits were considered together the distribution frequencies turned out to be discordant. That is, the geographical distribution of frequencies of one genetic trait did not match the geographical distribution of other traits. Clines for various genetic distributions were found to run in different directions. So, for example, blood group A occurs in relatively high frequency in East Asia and diminishes in frequency westward across the Eurasian landmass. The cline for skin color, on the other hand, has a north-south distribution.

Another way of looking at the problem of racial classification is to say that races exist *if and only if* the genetic distance (degree of variation) between a set of populations is lower than the amount of genetic variation between that set and other such sets of populations. If this *were* the case we *could* put valid boundaries around an intermediate unit (race) lying between the species as a whole and an individual population. The geneticist Richard Lewontin (1972, 1995) has claimed that we cannot do this. He argues that human populations are so polymorphic that the genetic variation within even any *one* population exceeds the genetic variation between it and any other such population.

Jean Hiernaux, a physical anthropologist using genetic data on several hundred African populations, provided data to support Lewontin's contention (1968). Hiernaux found that traditional classificatory schemes grouped some tribes that were widely separated from one another genetically. Conversely, other groups that had been thought to be only distantly related were, in fact, similar, but all of the then-existing taxonomic schemes failed to meet genetic criteria. Furthermore, no new classificatory system could be successfully substituted for the old ones. Variation among African populations tended to be continuous with almost no clustering of populations as real genetic units. When different trait classes were examined the data showed that these, too, did not conform to a racial model. There just were no neat categories beyond the level of the breeding population. More recently, Luigi Luca Cavalli-Sforza et al. (1994) have surveyed genetic data from around the world. Their data confirm Hiernaux's findings from Africa. It is important to note also that the data prove how *unnatural* the concept of race is even when what Coon referred to as "clinal populations" are eliminated from consideration. All populations are in fact clinal in respect to some trait. (Remember, the word "cline" refers to trait distributions *through* a *series* of populations.) Thus, attempts to find centers of racial homogeneity now appear fruitless.

The reasons for this lack of concordance must be explained. It is due, first of all, to the fact that humans among all animals are the most unpredictably mobile and the most widely distributed of species. Wherever humans migrate and encounter other populations, they interbreed. The human species bears little resemblance to those territorial animals among which breeding can only take place within a home range. Unlike the salamander, which, when forcibly removed from his quiet valley, will overcome almost all obstacles to return home and who will not breed while away, humans will mate wherever their migrations take them. Neither human females nor males display much territorial prejudice when it comes to sharing genes.

Migration and gene flow have spread human genes around the world in myriad ways. Successive migrations, conquests, absorptions, intermarriages, alliances, and extinctions of populations have produced a constant, never-ending shuffling of human genetic material.

Natural and sexual selection have also played their parts in the distribution of human genes, but the effects of these are often complicated and not well documented. As long ago as 1950 Coon, Garn, and Birdsell (all physical anthropologists) suggested a correlation between stature and climatic adaptation. These authors hypothesized that tall, thin individuals would have an advantage over short, fat individuals in hot climates. This was the case, they reasoned, because the tall-thin morphological configuration would allow for rapid loss of excess heat from the body, thus helping to avoid heat stress. The same reasoning predicts that short, heavyset individuals would be good conservers of heat and, therefore, should be found in extremely cold climates. In fact, the tallest thinnest populations are found among the Nilotic populations of East Africa and the Inuit peoples (Eskimos) are generally short and stocky.

However, Paul Baker, a physical anthropologist interested in physiological as well as genetic adaptation, has criticized the Coon, Garn, and Birdsell hypothesis:

> In contradiction to the original hypothesis put forth by Coon, Garn, and Birdsell, experimental work has thus far failed to show that a high SA/W (stature to weight) ratio provides any great physiological advantage to hot-desert dwellers. Beyond lowering water requirements because of a small body size, a high ratio has no appreciable effect on man's desert heat tolerance. Actually this could have been anticipated from a detailed knowledge of human physiology. The human body depends primarily on the cooling derived from sweat evaporation for maintaining thermal homeostasis in a hot desert. The hot dry air of the desert has enormous evaporative power and is apparently capable of evaporating sweat much more rapidly than the human body can produce it. Since the sweat production of the active man is related more closely to his fat-free body mass than it is to his surface area, the total cooling per unit of weight would be predicted to be very similar for men of quite different surface areas.

> Continuing on a theoretical basis, a high SA/W ratio may provide decided advantage to the man who must do physical exercise under hot, wet climatic conditions. When the air has a high moisture content combined with high temperatures, such as that in tropical forest regions, it no longer has the capacity to evaporate all the sweat produced by an active man. Under these conditions a significant proportion of the sweat will form water droplets and run off the man, providing no body cooling. Thus in a hot climate, the total surface area over which a given sweat production is spread will govern the amount of cooling derived. Since the amount of sweat produced by a man is governed by his weight of fat-free mass then, with activity held constant, increase in SA/W ratio should lower the heat strain on a man in the tropics. (Baker 1960, 6–7)

This quote shows how important it is to consider all possible environmental conditions when investigating the adaptive value of physiological traits, whether they be genetic or environmental.

Other explanations for the occurrence of particular morphological traits on the basis of natural selection have been offered in relation to skin color. Coon, Garn, and Birdsell (1950) suggested that dark skin is an adaptation to intense sunlight, which may produce severe burning with the risk of skin cancer. According to this theory, light-skinned people would tend to die of cancer before producing as many offspring as dark-skinned people living in the same environment. According to this theory, the heavy melanin (dark pigment) in the skin of Africans functions as a protective filter that cuts down the amount of ultraviolet rays absorbed by the skin. Citing distribution maps of gradients or clines for skin color Brace (1964) provided supporting data for this hypothesis. Brace, however, also suggests that dark skin is an original human trait congruent with the first appearance of humans on the African continent. This leads Brace to reverse the adaptation argument, hypothesizing that light skin probably resulted from the "probable mutation effect" that occurred with the relaxation of selection pressures as humans moved into less sunlit regions. Peter Post (personal communication, n.d.) has evidence that skin deeply pigmented with melanin is more sensitive to cold than light skin. Thus a reduction in melanin might also be an adaptation to colder climates.

The hypothesis concerning correlations between skin color and sunlight have not gone without challenge. Harold Blum (1961), a physiologist, suggested, for example, that sunlight-filtering devices in the skin are much more complicated than a simple deposition of dark pigment. He noted that protection against skin cancer involves changes in epithelial tissue. It has also been pointed out by other authors that correlations between sunlight and skin color have been based purely upon latitude, a simple north-south gradient, but that ultraviolet exposure is not completely contiguous with such a map.

Some authors question the sunlight-pigment hypothesis because they believe that dark-skinned Africans originated not in the open grasslands but deep in the tropical forest, where little sunlight penetrates. Coon (1965) is one of these. He believed that dark pigmentation is actually an adaptation to the damp chilly conditions found under the thick forest cover.

Whatever causes it, skin color worldwide is found to be relatively light in northern populations and gradually becomes darker in a north-south gradient. This is as true in the New World as in the Old, although southern New World populations are nowhere near as dark as Africans south of the Sahara. There may be a reasonable explanation for this difference. The migrations from Asia that led to the colonization of the Western hemisphere began sometime between 40,000 and 20,000 years ago. Thus, natural selection had much less time to operate in the New World than in Africa and Eurasia.

Similarities between the peoples of Africa and the geographically as well as genetically distant peoples of Melanesia (New Guinea and its island neighbors in the South Pacific) suggest parallel evolution toward darkly pigmented skin and woolly hair. These populations are so similar in external phenotypes (in their "packages") that it is difficult to distinguish, say, a Solomon Islander from an African. The apparent parallels between these populations led physical anthropologists in the prehuman-genetics period to group them under a single classificatory unit as Negroes. (The more cautious among these anthropologists classified Melanesians as Oceanic Negroes.) It now appears that this outward similarity is most likely *not* due to migration. This is shown by the great genetic differences that exist among these groups in such "content" traits as blood groups and other internal chemical factors. If the "package" is strikingly similar, the "contents" are not. It may be that a combination of sexual and natural selection has produced the black African and the genetically distinct black Melanesian.

Human diversity can also be related to natural selection operating on single gene traits. Among the most famous and most convincing of these (already discussed in the previous chapter in relation to adaptive polymorphism) is the high incidence of the hemoglobin Hbs allele in forested parts of West Africa. A relationship also appears to exist between certain other hemoglobin abnormalities and malaria. Among the best-studied are hemoglobin C, found in high frequency in Africa, and similar conditions known as thalassemia major and minor found along the Mediterranean littoral and in India, as well as in Southeast Asia. The high frequencies of these traits in certain widely distributed populations, taken by a past generation of physical anthropologists to indicate common racial ancestry, are now accepted as an indication of parallel adaptation.

Several hypotheses have been offered to relate other blood group distributions and other blood substances to natural selection. Many of these occur in frequencies that are best explained as cases of adaptive polymorphism. Livingstone (1958) in a summary article on the ABO(H) system has pointed out that individuals with different ABO phenotypes show differential susceptibility to a wide range of infectious and functional disorders.

There is another undoubtedly profound explanation for patterns of genetic variation among human groups. Here I refer to genetic drift. Drift, remember, represents accidental change, that is, change due to causes other than mutation, selection, sexual selection, or interbreeding. As noted in the previous chapter, genetic drift can occur when a population divides through the migration of a segment to some new territory or transgenerationally when gene frequencies of small populations are changed through such gratuitous events as floods, hurricanes, volcanic eruptions, earthquakes, and certain types of epidemics. The fertility of certain segments of populations may also be reduced significantly by social rules that produce differential access to mates or demand celibacy from certain categories of the population.

The effect of genetic drift has been thought by some experts on human variation to be slight. Garn, for example, offers the following summary criticism of its effectiveness:

> Drift, being purely chance, does not distinguish between adaptive and maladaptive genes. . . . Drift could account for some differences between adjacent populations, but not differences that are distributed in a regular way forming "clines." Only for perfectly neutral genes could drift operate alone to bring about major differences, and at the present time we are increasingly skeptical of such neutral genes. . . . (Garn 1961 and 1965, 97)

Garn was certainly right to insist that drift cannot be used to account for traits that are distributed clinally. On the other hand he failed to take note of the possibility that drift, *followed* by natural selection, could have profound effects on the development of genetic distance among once-related populations. Because natural selection operates independently of mutation and because evolution is an opportunistic process in which selection can only work on what actually exists, variations in gene frequency can be widened as selection operates on different populations living in different environments. The most common demographic pattern during most of the human species' evolution must have been one of small migratory or semi-migratory groups engaged in hunting and gathering. Such groups undoubtedly had a maximal population size in the low hundreds. Many were probably much smaller. We know from the study of living groups of this type that population growth and a concomitant reduction in natural resources leads to fission. Although considerable genetic contact between such groups may continue for some time, partial or even complete genetic isolation is an ultimate likelihood. Geographic factors such as separation by water or mountains and the exploitation of new and different environments are factors in the development of genetic isolation. Differentiation may also be enhanced by cultural change, such as linguistic variation, which occurs over time after contact between populations has been broken by the geographical factors just mentioned. This creates the perfect situation for drift. The smaller the original population, the more likely that fission will produce an unrepresentative division of that population's gene pool. Small size has the same effect on transgenerational changes, increasing the possibility that shifts in genetic structure will occur from one generation to the next.

The Nobel Prize winner Carlton Gkajdusek (personal communication, n.d.) believes that drift is so common a factor in the genetics of human populations that he cautions against assuming natural selection when island-dwelling populations are found to have similar gene frequencies. He suggests that such similarities are much more likely to be due to contact and interbreeding rather than to selection. In fact, he suggests that in most cases the action of natural selection on a series of island populations like those found in the South Pacific, even those whose parentage can be traced to a single ancestral group, would lead to increasing differentiation. He reasons that this would be the result first of drift and later of natural selection acting on altered gene frequencies.

Let us now return to the concept of race. I have already noted that the most generally accepted biological definition equates race with breeding population. I have also pointed out that this produces an unnecessary and potentially confusing redundancy of terms. It is equally important to note that such a definition renders it impossible for taxonomists to find and construct "race" as a real category between the species and the population. This is a rather bizarre situation because the concept of race as a biological category has always been offered for just such a purpose.

Carlton Coon attempted to get rid of the problems discussed in this chapter by disqualifying them. He believed that by doing away with so-called clinal populations we would be left with relatively pure centers, each characteristic of a particular race. The task is impossible because the mechanisms producing distance—mutation, selection, and drift—and those producing proximity—gene flow, interbreeding, and sometimes also selection—have acted in different ways in different places and times throughout human history. Some traits vary continuously, others, in discrete clusters. Those traits for which continuous variation (clinal) maps can be drawn do not all follow the same geographical lines or directions, so that no two frequency distribution maps are likely to be identical. These nonoverlapping maps plus the distribution of nonclinal traits create a jumble that renders an objective classification of races impossible. In spite of this Coon would have us return to an "ideal type" concept of race that only fits the outmoded typological notion of species and, in this case, subspecies (races).

At this point I realize that the reader may throw up his or her hands and say: "All well and good. I understand the problems involved in defining race, but my eyes tell me that races exist!" Lay people might even respond that they are more willing to trust intuition and eyesight than the technical ramblings of scientists. The scientist appears to have created an Alice-in-Wonderland situation in which things are not what they appear to be.

These feelings are misconceived. They add up to a funny-mirror distortion of reality, based on the assumption that external phenotypic traits have greater validity in classification than internal, invisible ones. Such a bias is simply wrong. In the past this assumption has led animal taxonomists up some very strange paths. Parallel evolution, in fact, has produced some amazing similarities between genetically distant species while differential environments acting on closely related organisms have produced rather spectacular differences. Who would guess, for example, that the guinea pig and the camel are closely related? Real taxonomic categories must be based on genetic similarity and distance. Additionally, the amount of variation present within any subdivision of our species may be submerged under selective perception and biased sampling. Most African Americans received *part* of their genetic heritage from *West* Africa. African Americans are, however, a mixed group with a considerable contribution of European genes and some American Indian genes as well. Given the history and geographic distribution of African Americans since their arrival in the New World, it is not logical to speak of them as constituting a single homogeneous population. Just as there are many "white" populations in the United States, there are many African American populations each with different gene frequencies.

The criteria used by the average person to categorize individuals as African Americans, or members of some other "race," are, more often than not, based on *social* rather than genetic identities. It is easy to identify an African American sociologically but extremely difficult to identify all sociologically defined African Americans on the basis of their phenotypes, even their skin color. The golfer, Tiger Woods, identifies himself as a mixture of African, Asian, white, and Amerindian background. Given the sociology of race in the United States, it is likely that many white and even black people in this country would simply classify him as African American.

If we were to build a racial classification on the basis of skin color and such other phenotypic traits as hair form, eye color, and facial shape, we would get a map differing from a map built from another set of phenotypic traits. There is no way of determining which of these maps would be "better" or "more accurate." In fact, none are "correct" or all are "correct." The only impossible situation would be for one of our maps to be correct while the others are in some sense false! The maps in question would reflect different aspects of human variation and not race.

In spite of all that you have just read in this chapter, consider the following quote from the 1975 edition of the Columbia Encyclopedia:

> Race, one of the group of populations constituting humanity. The differences among races are essentially biological and are marked by the hereditary transmission of physical characteristics. Genetically a race may be defined as a group with gene frequencies differing from those of other groups in the human species. . . . Nevertheless, *by limiting the criteria to such traits as skin pigmentation, color and form of hair, shape of head, and stature, and form of nose, most anthropologists agree on the existence of three relatively distinct groups: The Caucasoid, the Mongoloid, and the Negroid. . . .* (2263, italics mine)

Columbia Press published a new edition of their encyclopedia in 1993 with a revised definition of race. Unfortunately, the new wording, although it strongly cautions against confusing race with cultural categories, continues to ignore the difference between package and content. Thus even though genetic research had by 1993 provided ample evidence that black-skinned people in the Pacific region were genetically distant from African populations, the encyclopedia continues to group them with the so-called Negroid race. Although the new definition correctly points out that all human groups share more genes than differentiate them it continues to fall into the trap of typological thinking.

> One of the group of populations constituting humanity. The differences among races are essentially biological as marked by the hereditary transmission of physical characteristics. Generally a race may be defined as a group with gene frequencies different from those of the other groups in the human species. However the genes responsible for the hereditary differences between humans are few compared with the vast number of genes common to all human beings regardless of the race to which they belong. All human groups belong to the same species (*Homo sapiens*) and are mutually fertile. The term race is inappropriate when applied to national, re-

> ligious, geographic or linguistic, or cultural groups, nor can the biological criteria of race be equated with any mental characteristics such as intelligence, personality, or character. . . . Most scholars hold that there has been a common evolution for all races and that differentiation occurred relatively late in history. Even to classify humans on the basis of physiological traits is difficult, for the coexistence of races since earliest times through conquests, invasions, migrations, and mass depopulations has produced a heterogeneous world population. *Nevertheless, by limiting the criteria to such traits as skin pigmentation, color and form of hair, shape of head, stature, and the form of nose most anthropologists agree on the existence of three relatively distinct groups.* The Caucasoid found in Europe, North Africa, and the middle East to N. India . . . the Mongolian race, including most people of E. Asia and the Indians of the Americas. . . . The Negroid race is characterized by brown to brown-black skin, usually a long head form, varying stature, and thick everted lips. The hair is coarse, usually kinky, the eyes are dark, the nose bridge low and the nostrils broad. *To the Negroid race belong the people of Africa south of the Sahara, the pygmy groups of Indonesia, and the inhabitants of New Guinea and Melanesia. . . .* (2265, italics mine)

As my late colleague Morton Fried used to say: "You can't kill a bad idea!"

CHAPTER FOUR

PLAYING WITH FOSSILS

Carlton S. Coon's A Priori Theory of the Origin of Races

In this chapter I discuss the racial theories of Carlton S. Coon (1904–1981), whose work included research and writing in three of the four subfields of anthropology: physical anthropology, archeology, and ethnology (the study of living populations). To my knowledge, the only anthropological subject he did not cover (even at the beginning of his professional career) was the already highly specialized field of linguistics. Coon taught at Harvard University from 1927 to 1948. During that period, and well after retirement, he wrote both academic and popular books. Coon stands as an example of a man whose interpretations of the then-available evidence for human evolution were driven by a bad theory, the notion that blacks are inferior to whites in intelligence. This belief colored his interpretations of both fossil and living hominid forms and led him to speculations that were far from justified by the data.

So sure was Coon of his overall theory that it was to become the essential argument of two works, *The Origin of Races* and *The Living Races of Man,* books that, taken together, contain many internal contradictions. To my mind the case of Carlton Coon shows how an obsession concerning racial inferiority led to serious misinterpretations of the fossil record and the classification of modern human populations. Before I return to these works, however, I need to present a more general discussion of the theories of human evolution and racial origins that were current when Coon wrote to show how distorted Coon's ideas were even at the time they were proposed.

The history of hominid (human like) fossils began in Europe in the middle of the nineteenth century. Among the first specimens to be found were Neanderthals (originally classed as *Homo neanderthalensis* but now, as I have already noted, as a subspecies of modern humans, *Homo sapiens, neanderthalensis*). At the time the physical differences between ourselves and this find, although distinctive, were exaggerated by paleontologists. Although primitive in appearance, Neanderthals had brains larger than the average modern human. Somewhat later discoveries in Europe included the so-called Cromagnon man, identical to modern humans, and an enigmatic and massive jaw from Heidelberg that was to become the center of some controversy. (I will skip over the tale of the most controversial "fossil" of all, the now known-to-be-faked Piltdown man, a composite of an ape jaw and a modern human upper skull.) Other notable fossils included Steinheim man from Germany and Swanscombe man in England. These European finds were followed in China and Southeast Asia by the discovery, in the 1930s, of two fossils (now classed under the same genus and species as *Homo erectus* but at the time given two different genus and species names, *Sinanthropus pekinensis* and *Pithecanthropus erectus*). The African discovery, Rhodesian man, provided one of the earliest finds on that continent. We have already seen how Coon used the continental distribution of these finds in constructing the evolution of his racial categories. Suffice it to say here that the taxonomy of these (and other finds) presented a complicated and rather confusing picture of human fossil remains. Place names were the usual choice for each new fossil discovery. Additionally, those concerned with human origins tended to be very proprietary about their finds, often exaggerating the differences among the growing number of fossils. The result was an overly complicated nomenclature more often then not separating each new fossil into distinctive genus as well as species designations.

Well into the first half of twentieth century the focus of research in human paleontology was in Europe, where the first hominids were found, and, a bit later, in Asia. It was not until the 1940s that a tantalizing, very ancient, and controversial skull was discovered in South Africa by Raymond Dart. This find was given the name *Australopithecus africanus.* The fossil was controversial from the beginning because although apparently quite advanced toward modern humans one could argue that it was juvenile and, therefore, had not yet developed the more rugged features to be expected in an adult animal. Nonetheless, it attracted considerable attention. It was clearly quite ancient, at least one million years old, and exhibited such modern features as small teeth, in comparison to modern apes, as well as a much smaller facial projection than would be expected in fossil apes. Additionally, although small brained, a strikingly modern feature of this specimen is the placement of the foramen magnum (the hole under the skull through which the brain and spinal column are connected), located in a position that strongly supports the hypothesis of upright posture.

Primarily, it was not until the postwar years that the importance of the *Australopithecine* finds were clearly recognized with the discovery of a series of very early hominid specimens first in Kenya, by Robert Leaky and his associates, and much later in Ethiopia. Many of these were assigned to the genus *Australopithecus.* I will avoid most of the details of these discoveries and instead present a summary of current thinking about hominid evolution based on the most recent research. The first thing to note is the now-established fact that hominid evolution began in Africa and, even more importantly, that the first members of the genus *Homo* also appeared on that continent. Secondly, it is important to note that between two and three million years ago at least four species of hominids, some in the genus *Australopithecus* and, later, at least one in the genus *Homo,* lived in Africa, presenting a fertile field for evolutionary experimentation that eventually gave rise to modern humans in Africa and elsewhere. The first evidence of these early, African-based members of our own genus in Europe came only in the first quarter of the year 2000! Finally, as one might expect, as a greater number of finds accumulated a clear outline of the process of human development began to emerge. During the same period the tendency to name each new fossil after its place of origin gave way to a more rational taxonomic system based on anatomical similarities and evolutionary theory. What we are left with at present suggests the following sequence: Various forms of *Australopithecine* were precedent to the emergence of the first members of the genus *Homo,* and, for some time, both types of hominid coexisted in Africa, with early members of *Homo* migrating into Europe; the next stage saw the emergence and spreading throughout the Old World of larger brained and more culturally complex *erectus* forms, now all grouped into the genus and species *Homo erectus;* this was followed by a possibly archaic form of *Homo sapiens,* in turn giving rise to modern humans and the Neanderthals, the latter now classified as a subspecies of the genus *Homo* (*Homo sapiens, neanderthalensis* somewhat distinct from ourselves) and living contemporaneously with modern humans until their extinction about thirty to forty thousand years before the present.

If we now insert Carlton Coon into the picture we find that he was one of only two physical anthropologists, including then-contemporary workers in the field, to risk talking about "racial" evolution. The other, Franz Weidenreich, unlike Coon, however, never suggested that the different races emerged from totally distinct fossil lines, nor that one race emerged later than another, not to mention the possibility that this was responsible for the "backwardness" of one race in comparison to another. Given what it is possible to know from fossil evidence, it is clearly *not* possible to relate the evolution of the human species to the evolution of so-called racial differences, and no current physical anthropologist has attempted to do so. As we shall see below, Coon's reasoning on this subject was based on a series of shaky extrapolations from different kinds of evidence. In some cases Coon departed drastically from the timeline of human evolution presented by a large consensus of paleontologists most notably in his dating and classification of Rhodesian man as an *erectus* form when all others considered it a close relative of European Neanderthals. At other times he relied on contemporary African folklore to support a theory of race in Africa when no hard evidence existed at all. He needed to do these things in order to preserve his a priori assumptions about race, including his belief in the inferiority of contemporary Africans and their descendants in the New World. Let me now once again leave Coon behind for one last and short digression into theories about racial origin.

Two theories of racial origin, and hence human diversity, have haunted anthropology from its beginning. Even Charles Darwin got into the argument. One theory proposes an early division of what were to *become* humans into a number of races that evolved as separate lines from different hominid (prehuman) ancestors into modern *Homo sapiens.* The other theory proposes that modern *Homo sapiens* evolved only once and that "racial" variation is a relatively recent and superficial phenomenon. This was Darwin's position.

Scientists are not immune to the fact that moral and/or political considerations may color their approach to both theory and data. Both theories of racial origin have such moral and political implications. The argument that Coon supported for separate lines of racial origin from a presapient form has been used to claim inferiority of one or more races. The reasoning behind this notion requires the further hypothesis that some races evolved into the modern form before others and for this reason the "late comers" are "retarded." The second theory is generally, but not exclusively, the position of antiracists. As a recent and superficial phenomenon, racial differences cannot be taken as a sign of inferiority. Which theory more accurately describes evolutionary history is not yet completely clear, but a growing body of evidence strongly supports the late emergence of races as well as the superficial significance of so-called *racial differences.*

In the *London Review of Books* for January 27, 1994, Henry Gee reviewed three books concerned with human fossils and their place in evolution. Among the problems discussed in these books is the theory concerning early or late racial differentiation. Gee reminds us correctly that anthropological theories are no more exempt from political considerations than the theories of any other science. He goes on to say:

> Political correctness is a preoccupation of modern anthropology, and the history of physical anthropology as told by Trinkaus and Shipman [the two authors of one of the books under review] suggests that it is likely to be as distorting a filter as racism once was. . . .
> Trinkaus and Shipman recount the shameful treatment once meted out to the brilliant but politically naive Carlton Coon. Coon devoted his life to the study of racial variation, and reveled in racial diversity. . . .
>
> Coon's innocent but ill-judged antics were widely and publicly condemned as racist, notably by his younger and more idealistic colleague

> Sherwood Washburn, who felt that physical anthropology should be more socially relevant. The unfortunate result was its perversion from a science into a series of excuses to justify a particular moral stance.

It should be clear from what I have said above that I share Gee's concerns about politics and science. I have never been comfortable with the way some anthropologists treated Coon's work, particularly the attempt to censure him publicly at the annual meetings of the American Anthropological Association in 1966. Although I am convinced that on scientific grounds Coon was wrong, I was upset about the incident because many of those present calling for sanctions against Coon clearly had read neither of the two books in question. But what Gee appears to ignore in his defense of Coon is that the political views of scientists can cut two ways. There *is* enough material available in Coon's own publications to show that it was his racist attitudes that drove his interpretation of data and not the reverse. While I personally did not vote to censure Coon, I have published criticisms of his conclusions. I base these, as far as is possible, on a critical look at Coon's data and the logic of his arguments. Before demonstrating this, however, I have to back up somewhat and discuss what was known of human evolution at the time of Coon's two books on race as well as his major hypothesis concerning the evolution of separate racial lines from presapient (*erectus*) ancestors as it was proposed by the man who most influenced Coon in this respect, Franz Weidenreich.

At the time Weidenreich wrote (in the first quarter of the twentieth century), the existing *erectus* fossils were divided into their own genera and species. The most notable among them were the Chinese form, then known as *Sinanthropus pekinensis,* the Javanese form, then known as *Pithecanthropus erectus,* and the assumed *erectus* lower jaw from Europe, the so-called Heidelberg man. Weidenreich, realizing that long periods of isolation between the various continental forms of *erectus* would have given rise to speciation, suggested that the parallel development undergone by these and other *erectus* fossils toward modern *sapiens* was insured by constant gene flow at all points during the long process of evolution. Weidenreich never suggested that some races evolved into modern humans later than others. It was Carlton Coon who argued that the boundary between presapients and sapients was crossed at different times with the overt implication that some races were inferior to others.

When Coon published the two books in question, the available material on human evolution was relatively scarce compared to the present. For some scholars a series of fossils from Africa suggested an early appearance of prehuman forms in Africa, but there was not yet the plethora of fossils found in recent years that point to human origins (*Homo* if not *sapiens*) on that continent. At the time the only examples of the genus *Homo,* to be discussed below, were from Asia and Europe.

The purpose of Coon's two books on race was to trace the fossil origins of five human subdivisions said to represent separate lines of evolution toward modern *Homo sapiens.* Once established, these five lines (Congoid, Caucasoid, Capoid [African Bushmen], Mongoloid, Australoid) were hypothesized to be temporally continuous and relatively separate spatially. Although each line changes through time, each maintains biological integrity. According to Coon, each of these five lines passed their evolutionary history in much the same way. All were destined to evolve from the early hominid types into modern *Homo sapiens* through the intermediate step of the *pithecanthropine* (*erectus*) fossils. All were successful in attaining the sapient grade, but some were to reach it before others.

For Coon (1962) races became differentiated from one another through the process of specific adaptation in which populations are modified to fit local environmental conditions. The species on the whole, on the other hand, evolves as a unit. This occurs when adaptations of a general sort spread from population to population through gene flow and then gain rapidly in frequency through natural selection, which favors them in all environments. Thus, for Coon, the epicanthic fold (found primarily in Asian populations) is a specialized adaptation limited to one type of environment, while the development of a better brain has general selective value. As long as gene flow occurs, a species can evolve as a whole, but subunits can become differentiated from one another through isolation. Coon saw the different races crossing the various evolutionary levels at different times. For him the crucial threshold was between the *erectus* stage and modern *Homo sapiens.* Coon said that it occurred early in Mongoloids and Caucasians, late in Congoids, as he preferred to call black Africans, and latest in the Australian aborigine. The unstated assumption was that most, if not all, of the generalized positive mutations occurred in one, or at most two, of the five lines of evolution, and that they got to the other lines primarily through gene flow.

Most of Coon's critics found this five-line theory untenable. It is difficult to envisage a parallel evolution of five species toward the same product. Of course such a parallel evolution can occur, particularly when environmental restrictions on development are quite rigid and when the evolving types come from a common stock. Parallel evolution occurs even in unrelated species when they are subject to the same highly restrictive environmental conditions. Thus, many desert-dwelling animals of only distant common lineage have developed a series of similar (i.e., parallel) physiological and behavioral adaptations to high temperatures, low humidity, and a scarce water supply. Such adaptations involve internal water conservation through accommodations in kidney structure. Desert plants from widely divergent lineages also display certain parallel developments, all of which contribute to efficient water conservation. These parallels exist, however, as a specific set of adaptations to a very restrictive environment. Such animals and plants also diverge in other morphological and physiological features because each species occupies its particular environmental niche within the larger desert environment. They are different also (and this is very important) because they have different genetic histories. What are possible adaptations for some species are impossible for others because of the equipment they brought to the desert environment in the first place. No biologist expects species with parallel adaptations of this sort to be capable of interbreeding. They are too different genetically. Their similarities in structure are usually controlled by different sets of genes.

Coon's parallel evolution is of a different sort. The five units concerned evolve toward the same general end through a considerable time span and maintain their interfertility. Hasty critics were quick to attack Coon for this. Coon's model, as it has been offered thus far, is perfectly reasonable from an evolutionary point of view. The five races begin as a single species, perhaps one "race"; they spread out and diverge, but they are held together by a considerable amount of gene flow. Centrifugal forces are balanced by the centripetal force of gene flow, which acts not only to keep the species intact but to pull lagging populations along with the general evolutionary trend.

While this part of Coon's theory is acceptable, facts suggest that it is wrong. Wrong because the human species does not stay put to the degree that some other animal species do, and wrong because human genes flow faster and in greater numbers than Coon would have it. Humans are a promiscuous species. For Coon's theory to work, the process requires just enough gene flow to maintain species integrity, but not too much or the races would blend too much. It also requires geographically static populations. The mobility of prehistoric humans was probably limited in comparison to historic times: Means of transport were limited to foot travel and human populations were hemmed in by environmental barriers. At the same time, however, hunting populations (all humans were hunters and gatherers until approximately 10,000 years ago) are mobile by necessity. They must follow the game wherever it goes. Over the course of several hundred thousand years, game animals have been pushed hither and yon by environmental variations, including glaciers and progressively wetter or drier conditions, and possibly also by the effects of hunting itself. Human hunters have been forced to follow the game at least since the species acquired culture. In addition hunting populations are bound to be rather small. Such groups are perfect vessels for the operation of genetic drift. Rapid evolution (and the emergence of *Homo sapiens* in evolutionary terms was indeed rapid) is most likely to occur in small populations connected intermittently through gene flow. Such populations act as evolutionary laboratories. In such a situation no single line is the source of major adaptive changes.

There is another problem in Coon's theory. This is the assumption that there have always been the same five lines of human evolution and that these lines have a historical depth and continuity that carry them back to the near beginning of humans as a species. Human evolution has been more complicated, with frequent population mergers and splittings. All the forces of evolution—selection, interbreeding, and drift—must have played a continual role in the differentiation and remerging of human groups.

The major problem with Coon's theory, however, is his belief (for which there is no evidence) that certain races became fully human before others.

Here his a priori racist thinking most clearly drives his interpretation of data. If all races had a common and recent (recent in terms of thousands of years, it must be understood) origin, he asks rhetorically, how could some people, for example the Australian aborigines, still live in a manner comparable to that of Europeans 100,000 years ago? Coon answers his own question by claiming that the ancestors of these groups and the soon-to-be advanced Europeans parted company in remote antiquity. Otherwise, he claims, Australian culture would have had to regress rapidly, a process for which there is no archeological evidence. Coon also argues for the antiquity of his separate lines from linguistic evidence. If the existing races had been rooted in a single population only a few thousand years ago (it is more like 50,000 years ago) linguistic divergence today could not be so pronounced. No one need argue that human divergence is a recent phenomenon to counter Coon's theory. What he does here is set up a straw man. There is no doubt that as the human species spread out over the face of the earth (archeological evidence shows that this process must have occurred early in human prehistory) groups diverged culturally and linguistically and, to a lesser degree, physically. What can be called into question is the immutability of Coon's five racial groups and what Coon sees as a short timeline for cultural and linguistic change to occur. Both of these can, in fact, be quite rapid. In addition, cultural loss as well as technological advances are known to have occurred during several periods of human prehistory. There are documented cases of rapid cultural devolution with a return to more technologically primitive conditions. Such was the case for the San (bushmen) of South Africa, who were hounded out of their natural habitat by more technologically advanced Bantu populations as recently as the end of the sixteenth or seventeenth centuries. Refugee populations ("losers") are frequently found to have "regressed" culturally to more primitive conditions than those under which they had previously lived. There is no reason at all to assume correlations between gene flow and the flow of culture. People "accept" genes without accepting ideas just as they may accept ideas without accepting new genes.

Coon assumed that the biological differentiation he found among human groups is just too great to be of recent origin. Interestingly, Darwin thought otherwise and tested his notion by studying emotional expression around the world. This led to the publication of his *Expression of the Emotions in Man and Animals* in 1872. In that book Darwin claimed to have proved a monogenic origin for humans and a recent evolution of the living races. He found racial differences among human groups to be quite superficial in nature. It is interesting that Coon took an opposite view, for he is the same physical anthropologist who defended a case for similarity between ourselves and Neanderthals.

Coon had the bad habit of manipulating degrees of difference depending upon what argument he was defending. His subjective and preconceived notions about race got in the way of his supposed scientific objectivity. At times Coon exaggerated differences among human groups (races); at other times he ignored them (Neanderthal populations versus modern humans). If we look at a species considerably younger than humans, namely domestic dogs, we find a truly amazing degree of polymorphism. Certainly the pygmy is closer to the Englishman than the Pekinese is to the Great Dane. True, the differentiation in dogs, a domestic animal, took place rapidly through artificial selection imposed by humans, but assortative mating, particularly sexual selection, may well have had an equivalent effect on human differentiation. Add the effects of natural selection and drift to this and we need not call up great antiquity to account for the range of existing human types.

Coon argued that some contemporary races preserve a high frequency of archaic traits that can be traced back directly to the *erectus* stage. This is taken as evidence for late emergence into the sapient form. This conclusion is suspect for a host of reasons. First, if selection pressures have been such that "archaic" traits such as large jaws and big teeth tend to be preserved in some populations, then no amount of time would eliminate them. Second, since there is a good deal of physical variation within all human populations, drift alone could rapidly shift the average type of physical appearance. This shift could either be in the direction of "progressive," "more sapient" features or what Coon would subjectively label more archaic or *erectus* features. Furthermore there is no guarantee that the gene complexes that cause "*erectus*" features in modern populations are the same ones that produced *analogous* traits in *Homo erectus* thousands of years ago. Thus the use of so-called archaic traits to work out the history of contemporary populations is a very unsound methodology, and the use of the term "archaic" loads the dice unfairly.

In *The Origin of Races* Coon traces what was known of fossil humans from Africa, Europe, Asia, and the Pacific area at the time (the 1960s). Fossils found in Europe were taken as Caucasoid; those from Asia as Mongoloid, etc. Coon stated that the principal differences between Negritos (Pacific pygmies, Oceanic Negroids, and Australoids) were body size and hair form. Thus, by ignoring other phenotypic traits Coon could claim that these three populations evolved from a common local ancestor. That these groups differ only in these two traits was a surprise to most physical anthropologists familiar with the complexities of genetics, but Coon seemed to feel that differences based on single genes, such as blood groups, were of little significance (at least for this geographic area and in this book). Once again, even with the limited knowledge of the time Coon was interpreting the available data through the filter of a priori racist assumptions. I know of no other physical anthropologist in the 1960s who would have dared to offer the same conclusions.

Coon looked at the known fossil sequence, which he believed led to the modern Australoid. This began with the fossil *erectus* found in Java, through *Solo* man, to *Wadjak* man. The latter is given the status of a full *Homo sapiens* in brain size but is categorized as "flat faced" and "broad nosed." The jawbone displays a chin (absent in *erectus* fossils), but Coon admits that this may or may not be a diagnostic feature of *Homo sapiens* and is "massive." Other skulls, all *Homo sapiens* (Keilor, Talgai, and Cohuna) tie Australoids into a series of Pleistocene populations in Java. The fact that the aborigine skull is massive in many of its features was taken as evidence that the Australian is still in the process of "sloughing off" a series of genetic traits that link him with *erectus*.

In his discussion of Mongoloids Coon listed seventeen features they have in common with *Sinanthropus*, the Asian mainland representative of the *erectus* grade. These traits were taken from Franz Weidenreich's model of racial evolution, which includes parallel development dependent on a constant and high degree of gene flow among the various evolving racial lines. Many of the traits listed by Coon are related to the chewing apparatus and are therefore implicated in the type of diet eaten by these populations. Such traits may or may not have a genetic base. It is possible that some jaw and muscle development is the result of physiological adaptation (response to stress) rather than genetics. Though Coon makes no mention of it, many of the traits he lists as Asian in origin, are, in fact, found in European and African fossils and some even in certain living non-Asiatic groups. Besides, Coon made no attempt to place these seventeen traits in the context of an overall morphological pattern. Thus, he never provided a means for determining the general importance of the traits in the determination of genetic (better morphological) continuity among the world's populations distributed, as they are, both temporally and geographically.

In dealing with Caucasians in *The Origin of Races,* Coon suggested that the lack of *erectus* fossils in Europe (he wrote before the European *erectus* finds were uncovered) points to their origin on some other continent, perhaps western Asia. Coon read the archeological evidence (tool types) to say that the original Caucasian populations had spread over western Asia, Africa, and Europe by the Middle Pleistocene. In the Upper Pleistocene, however, cultural differentiation took place, with the divergence of tool types on the African continent. Skeletons found in Palestine, Lebanon, Iraq, and Uzbekistan are all typed by Coon as Caucasian, while the contemporary African material is classed as racially different. The placement of early Caucasoid peoples in Western Asia, he said, gave them a genetic advantage, for they lay on a path of gene flow from all the Old World continental landmasses. For Coon the most likely populations to have contributed their genes to Caucasians were the Australoids in India, the Mongoloids, the Capoid populations of North Africa (these people are now limited to Southern Africa) and perhaps even Congoids through contact with southern Arabia.

The opportunity for Mongoloids to benefit from gene flow was contrasted by Coon to that of the Caucasians. He blames genetic isolation for the Mongoloid's extreme racial peculiarities. Europeans, on the other hand, were in a position to accept genes directly from the three other races, process them through natural selection (both climatological and cultural), and pass the putative benefits back to peripheral populations.

Thus, for Coon, the Mongoloid owed his sapient nature to his own mutations, but the European–West Asiatics were in a position to capitalize favorably from advantageous mutations from more than one racial group. The gift of sapient status could then be passed on to less fortunate populations (presumably Congoids and Capoids). Such mixture makes the Caucasoids the "least pure" of all human races, but this is taken as an advantage from the evolutionary perspective. If there ever were a deus ex machina to explain the origin of races, this is it.

This marvelous machine, however, is (without a word of comment) directly contradicted in Coon's subsequent book, *The Living Races of Man* (1965). Here we are told that while the Caucasians *are* physically highly variable, most gene flow was apparently out of the European area. "Much has been written about the influence of Negroid infiltration into Mediterranean countries and about Mongolian genetic penetration in Eastern and Central Europe. Both have been exaggerated. . . . The Moors, that is the Arabs and Berbers, occupied much of Spain and Portugal for seven centuries and Arabs also held Sicily for a while. Arabs and Berbers are themselves Caucasoids, but they brought a number of Negro slaves with them" (Coon 1965, 66).

While there might be a small grain of truth in this, overall it is pure fantasy. The time scale discussed here is so short as to be of little value in understanding the dynamics of population mixture in the Mediterranean basin over the course of many millennia. In addition, the Arabs must have been a mixed population in the Middle East well before the historical period discussed here by Coon. Finally, what is most unimaginable in this scenario is the static picture it proposes. Genes do not flow in only one direction nor do they require migrations to flow. Adjacent populations also mate with one another. Here, if Coon is to have his way, we have a region with a tremendously long period of contact between a complex series of populations characterized by a wide range of cultural identity, existing around a navigable lake that was fully exploited for trade and conquest, giving rise to a one-way, or predominantly one-way, flow of genes. This is preposterous! Logic demands that what really occurred was merging that resulted from two-way gene flow *and* differentiation that resulted from both drift and natural selection. Circum-Mediterranean populations share many genes. They are, in fact, what Coon refers to elsewhere as "clinal populations."

Turning to fossil material, Coon delineated a series for the Caucasian line going back to the Maur (Heidelberg) jaw, which has, he said, characteristics similar to both African and Australoid groups but which is divergent from the Mongoloids. The turning point for Caucasians, said

Coon, is the Steinheim skull that is dated to the Mindel-Riss interglacial period 250,000 years before the present or 110,000 years younger than *Sinanthropus* (the Chinese *erectus*). Coon stressed that Steinheim's cranial capacity (1150–1175 cc.) does not differentiate it from the Javanese and Chinese *erectus* line but that other features of the skull do show advances toward modern *Homo sapiens* over the *erectus* grade. This, he said, is particularly noticeable in the occiput (the base of the skull), which is well rounded, a modern trait. The forehead is low but quite steep; the mastoids are small (which, by the way, is an ape-like characteristic); and the side walls are parallel.

The next skull in Coon's series was found at Swanscombe, England. Its cranial capacity is given at between 1275 and 1325 cc. This puts it in the modern European female range (the average woman has a smaller brain than the average man not because she is less intelligent but because, on the average, she is smaller than the average man). Measurements of this skull for breadth and height closely resemble modern Caucasoids. Coon went on to say of this fossil: "There has been a great deal of speculation about Swanscombe's face, but because Steinheim has a face, and because the threshold between *Homo erectus* and *Homo sapiens* lies in the brain and not the face it is unnecessary" (Coon 1962, 495).

The involuted reasoning in this quote is a milestone in physical anthropological analysis. My translation would read: It is brain size that determines the grade and not the face, but the face of Steinheim is sapient and not *erectus,* therefore it is likely that the face of Swanscombe is also sapient! This analysis stands in strange contrast to that given to the Rhodesian skull (see below), which is classed by Coon as *erectus* and by many, if not all other anthropologists, as *Neanderthaloid.*

The Swanscombe find as well as another fossil from Fontechevade in France are also offered as evidence for early sapient emergence in Europe with the additional assumption that these early populations were Caucasoid.

In *The Origin of Races* Coon devoted a chapter to Africa, which, he claims, is the home of the Capoid and Congoid subspecies. He admitted that the evidence for Congoid origins is so scarce that it is difficult to interpret, but he nonetheless builds a sequence by marrying existing fossil material to speculation about the effect of natural selection on African populations.

Most physical anthropologists *would dare* hazard a guess about the race of a single skull, particularly if it were of a modern human, but most would much rather have a complete skeleton and would be yet considerably happier with a fairly large-sample skeletal population from which they could draw a set of average measurements. As I have emphasized before, human populations are very polymorphic, and it is, therefore, difficult to draw sound conclusions from single specimens. Yet Coon was able to extract the following information from the incomplete Rhodesian skull (it lacks a lower jaw).

> In the frontal index of facial flatness and in the simotic index (reflecting the archings of the nasal bones at their root), the Rhodesian skull falls within the Caucasoid range, and in the third or rhinal index of facial flatness it resembles the ancient Egyptians and approaches the means of published series of living Negroes. In the fourth or premaxillar index of facial flatness it leaves all modern populations far behind it. In other words, this face is Caucasoid in its upper portion, Congoid in the middle and virtually pongid below. On the whole the face is mostly Negro. . . .
>
> The tibia, which was found at the bottom of the cave with the skull, resembles that of a modern Negro in all essential details. (1962, 626)

In addition, we are told that Asselar man, another fossil from West Africa, was a Negro from the neck down!

Coon needed to have the Capoids originate in North Africa so a native myth is cited as evidence for this geographic placement. The Riffians are said to have a vivid image of their predecessors: "They had the ability to transform themselves: a *thamza* [female] could turn into a bewitchingly beautiful Berber damsel, and an *amziv* [male] into a Negro. *Obviously then, they were, in their natural forms, neither Caucasoid nor Negro*" (ibid., 601–2; final italics mine).

And then Coon refers back to fossil material: "If the Ternefine-Tangier folk were not the ancestors of the Bushmen, they were a sixth subspecies that uniquely died without modern descendants, and the Bushmen would have no discernible ancestors" (ibid., 602).

Turning to the effects of natural selection on racial origin, Coon suggests on page 589 that the Negro originated in the Savannah lands at the edge of the tropical forest. Black skin is offered as an adaptation to the possible deleterious effects of strong sunlight. But in *The Living Races of Man,* published three years later, he offered an entirely different theory, the one that I have already mentioned above: "We suggest that one function of deep pigmentation in Negroes is to keep them warm" (1965, 233). Thus, the Negro homeland becomes the deep forest with its protective cover. The forest climate is damp, and in the rainy season the combination of relatively low temperature and high humidity can produce chills. Dark skin, which would absorb more heat from the hearth than light skin, provides an advantage even in an environment in which the sun's rays rarely penetrate to the forest floor.

The contradictions in the two books went largely unnoticed in reviews, yet in *Origin,* Coon had the Pygmies developing out of the Negro with a *penetration* of the tropical forest. In the *Living Races,* Pygmies emerge as the original black population of Africa. On page 100 we are told that the Pygmies were probably the original inhabitants of West Africa, and on page 105: "We can generalize that in every measurable or observable character known the Pygmies stand at one extreme, the African Caucasoids, at another, and the Negroes in between." On page 123 we are told that an analysis of the crania indicates that Negroes gravitate between Mediterraneans, Caucasoids, and Pygmies: "The evidence suggests that the Negroes are not a primary subspecies but rather a product of mixture between invading Caucasoids and Pygmies."

But elsewhere Coon had claimed that mixtures do not produce races and that most of the gene flow was the other way (from Pygmies to Negroes), and he never includes Pygmies as a sixth racial group. Not only do the two books contradict one another, they tend to garble Coon's whole theory of racial origins. Since they represent in Coon's own words two parts of a single work, it is difficult to ascribe differences between them to changes in the author's thinking.

The reader who accepts Coon's rule that clinal populations need to be eliminated from consideration when constructing races will be surprised to find out that "today the indigenous population of Africa is mostly Clinal. In the Sudan and East Africa, Caucasoids shade into Negroids; and telltale pockets of partly Capoid peoples survive in the Sahara and along its northern fringes" (1965, 84). If we follow this we are forced to eliminate "Negroes" from any consideration of races for

Coon had already informed us that clinal populations do not count. But if this cline extends downward into Africa surely it would have to extend upward into Europe as well. Whoops! This relegates Caucasoids to the status of a cline as well. This would leave us with only two "real" races: the Mongoloid and the Australoid.

In *Origin* Coon has genes flowing into Europe; in the *Living Races* he has them flowing out of Europe. Perhaps Coon wanted us to read these contradictions out of existence by adjusting them to temporal changes in gene flow. Thus, early in fossil development there was flow into Europe, and later, a flow out of Europe into Africa; but, even if we were to grant this, why the change in direction? The whole argument would destroy Coon's theory of races as entities, at least for Negroes and Caucasoids.

Coon admitted that the fossil evidence he called forth to support his claim for the late emergence of the *sapiens* grade in Africa was poor. The Chellean–3 skull from Olduvai gorge, which is classed by most physical anthropologists (including Coon) as *Homo erectus* and dated contemporaneously with both Heidelberg and *Sinanthropus,* was offered as a possible ancestor for both Caucasoids and Congoids. Coon rested his argument on the Rhodesian material, not only for the divergence of the Congoid line but for the late evolution of this line from the *erectus* to the *sapiens* grade. As I have already indicated, most physical anthropologists place Rhodesian man in the Neanderthaloid group, above *erectus* on the phylogenetic scale of hominid evolution. None to my knowledge would care to attribute any racial affinity to it. Coon's interpretation of this skull is at best highly dubious.

A skull from Cape Flats, South Africa, is said by Coon to follow the Rhodesian pattern in which Caucasoid and Negro features are combined. This and another fossil, the Border Cave skull, classified as Australoid by South African physical anthropologists, are taken as Congoid by Coon, who saw South Africa as a cul-de-sac for the remains of early Congoids. He asserted (1962, 633) "Whether or not a local race of Negroes evolved in South Africa before the ancestors of the Bushmen arrived has little to do with the origin of the Congoids as a subspecies. . . ." Coon stated flatly that the Negro originated in West Africa, an area from which, he himself admits, there is not a scrap of evidence. How convenient.

The oldest Negro fossil is that of Asselar man, found near Timbuktu. This, I must remind the reader, is the post-pleistocene specimen "wholly Negro from the neck down." Coon cited other skeletal material, but it dates from such a recent period that it adds little substance to his argument.

In *Origin* Coon suggested that the Negro, who was originally adapted for savanna living, developed adaptations to the forest environment through intermixture with Pygmy groups. This is rank speculation, there being no evidence for the borrowing of a forest genotype on the part of Africans from some other group, nor for the origin of Negroes through the admixture of Pygmies and Caucasoids (Coon's second theory of Negro origins). And after all this Coon was faced with the self-stated problem: Who are the Pygmies? His solution is yet another flight of fancy. He agrees with the late Father Gusinde (a missionary priest and cultural anthropologist) that Pygmies are descendants of old, pre-Hamitic, pre-Capoid populations of the African Savannas who were driven into the forest by drought.

Coon's first theory is that the original proto-Negro, proto-Pygmy group crossed with true Pygmies to produce the modern Negro (1962, 66). It must be stressed yet again that we know virtually nothing of the origin and development of recent African populations. We also know nothing of early differentiation. If (as is certainly possible) black African populations emerged late, this would reflect only one case of the kind of divergence that can take place within a polytypic species over and over again through the operation of the genetic processes described in the second chapter of this book. A possible late emergence of the black African, however, has nothing to do with Coon's hypothetical late crossing of the line between *erectus* and *sapiens.* For racial, or better, populational differentiation can take place within a species at any stage of evolution. In *The Living Races* Coon suggests that Negro origins can be tied to a cross between Caucasians and Pygmies late in the biological history of the human species. The arguments, therefore, that Negroes are somehow inferior to Caucasoids, which have been employed by some of Coon's *followers,* make no sense on any ground.

I have one other major criticism of Coon's work. The photographs of members of different human populations presented in *The Living Races* have little or no scientific value for comparative purposes. They lack control for differences in age, sex, dress, hairstyle, facial expression, or even angle of photograph. Furthermore, they lend support to typological arguments because they are open to interpretation as "type specimens."

We are shown, for example, four Basque males (plates 77–78 a, b, c) to show "a wide range of facial features"; three Tuaregs (plates 107 a, b, c)—"One of the expected lean and aquiline type" and "two other Tuaregs who look like ordinary Berbers." But plate 109, we are told, is a Berber with "Bushman-like features." Such photographs tell us nothing of real variation in physical type within each of these groups. One might reply that Coon has done his best in assembling 128 plates covering a wide range of peoples. But since these pictures have no scientific purpose, why include them at all? The captions are also very misleading. Many refer to ethnic or linguistic identity and give not the slightest clue to genetic affinity. Thus under circumpolar peoples we find (3a) a Zyrian, with the caption "The Zyrians are Finnish people hunting and herding reindeer in the forests of northern European Russia" and (3b) a Vogul, with the caption "The Voguls are Ugrian-speaking people of the Obi River country who live principally by fishing." Plate 11c is a "Siberian woman with a Ukrainian father and an Eskimo mother." Coon tells us "She looks completely European." Because of her haircut and her general facial structure, she looks Japanese to me, but no matter. Plate 33 is "A Japanese nobleman of aristocratic facial type." There is, however, no mention in the text of differential physical types among Japanese social classes. In plate 38 we are shown a "Tibetan lama with curly hair," but we are never told the frequency of this trait among Tibetan lamas, or any other Tibetans for that matter. Plate 43 represents three Nagas from northeasternmost India and adjacent parts of Burma. "The Nagas resemble American Indians in appearance." To me the man in the picture does, the women don't—but which American Indians, and how representative are these Nagas?

The most misleading, and I must say scandalous, set of pictures, however, appears in *Origin,* plate 32, which shows an Australian aborigine woman and a Chinese man with the following caption: "The Alpha and Omega of *Homo sapiens:* an Australian aboriginal woman with a cranial capacity of under 1000 cc. (Topsy a Tiwi); and a Chinese sage with a brain nearly twice that size (Dr. Li Chi, the renowned archeologist and director of Academia Sinica)." The inference is that Dr. Li Chi's intelligence, of which we have no measure, is vastly superior to that of Topsy (no measure here either) and that the difference is due to brain size. Topsy could be a genius for all we know. Correlations, within the normal range, between brain size and intelligence do not exist. Females often have smaller brains than men not because they are less intelligent but because they are smaller on the average than men. Finally, although Coon gives us an estimation, we don't really know the cranial capacity of either Topsy or Dr. Li Chi. This picture with its accompanying caption tells us more about Professor Coon's preconceived notions than about either person in the photos or the mechanism of intelligence in the human species. It certainly has nothing to do with the origin of races, except that for Coon the Australian aborigine is the result of late (and, therefore, retarded) evolution.

Many of Coon's critics, in their attempt to be fair, have stated that *Origin* is an excellent compendium of fossil humans and prehumans, and so it was in its time. But the book was not written as a catalog. It has an argument to offer, one that is deeply flawed.

CHAPTER FIVE

RACE AND IQ

Arthur R. Jensen and Cyril Burt

This chapter deals primarily with race and IQ in the work of Arthur Jensen, the author of a 1969 report on race and IQ that was essentially an attack on the governmental program known as Project Head Start. The program's goal was to help children from poor neighborhoods prepare for their entry into the regular school system through attendance at free government-supported preschools. The assumption behind Head Start was that the children of the poor suffered a learning deficit in their early formative years due to an impoverished intellectual environment. For those who believed that IQ and, therefore, performance was hereditary, Head Start was seen as a waste of federal monies.

Jensen is an educational psychologist specializing in psychological statistics who, after many years as a professor at Teacher's College, Columbia University, moved to the University of California at Berkeley. His highly controversial article "How Much Can We Boost IQ and Scholastic Achievement?," published in the *Harvard Educational Review* in 1969, made a case for the preponderance of heredity in the production of intelligence as measured by IQ tests, and an average genetic deficit in IQ among people of black ancestry when compared to whites. Although the argument had been made before, Jensen's article drew a vast amount of positive attention from the press and among some educators and strong criticisms from many, but by no means all, professional psychologists and anthropologists. It is important to note that the "Jensen Report" came shortly after the Supreme Court decision banning segregation in public schools and the successes of the civil rights movement to desegregate schools in the South. Therefore, it should come at no surprise that Jensen's conclusions were seized upon immediately by those who opposed remedial educational programs, such as Project Head Start, for young poor children and, in particular, poor black children. In a nutshell their argument was: If, as Jensen has proved, IQ is largely hereditary, it is a waste of money and time to develop and pursue programs for children in order to enhance their intelligence. Because even today this article stands as a model for those who continue to believe the IQ argument concerning race, this chapter will focus on its major shortcomings. Later work concerning group differences and IQ will be taken up in chapter eight, which is devoted to Herrnstein and Murray's *The Bell Curve.*

In discussions concerning hereditary group differences in IQ, race has not always been the crucial variable. In Great Britain, for example, the focus has been on class rather than race. The man most associated with modern work on class and IQ in Britain, and who had a significant impact on Jensen's methods of research, was Cyril Burt. Burt attempted to prove that heredity played a major role in intelligence by studying identical twins reared apart. Because such twins are genetically identical, any differences in IQ found among them must be due to environment. However, because no one has ever argued that genetics plays the *only* role in the determination of IQ, such studies are putatively used to determine the proportional contributions of genetics and environment to a trait that varies among populations with different genetic profiles and brought up under different environmental conditions. Since at the time that Cyril Burt did his work class and not race was a major concern in Britain, he set out to prove that IQ was the *major* variable in *class* differences in intelligence.

This idea was not new with Burt. It was first proposed in the middle of the nineteenth century by Darwin's brilliant cousin, one of the founders of mathematical statistics, Francis Galton. Galton warned that class differentials in fertility, with the lower class having more children than the upper classes, would inevitably produce a gradual decrease in the average IQ of the entire British population. Burt represented a modern version of Galton's hypothesis and provided what he claimed was solid evidence of the phenomenon that would bring about the decline in IQ predicted by his predecessor.

Before I criticize Jensen's and Burt's work in detail let me turn to the concept of race and its purported relation to behavioral traits including—but not exclusive to—IQ. As we have already seen, the concept of race is often confused with ethnic, cultural, or religious identity. People speak of the French race, the Irish race, or the Jewish race. In societies where racism is current, individuals of mixed ancestry are usually assimilated into whichever part of their ancestry is downgraded by society. Thus, in the United States even individuals who are phenotypically white may be classed as black if it is known that they have even a small degree of black ancestry. It is fair to say, therefore, that even if race is a false concept in biology it is *real* from a sociological perspective. When members of a society classify an individual by race then that person *is by definition* a member of that race!

Racial identity is by no means a neutral concept. Wherever used it implies superiority or inferiority. Which "racial" groups are esteemed or denigrated is determined by subjective factors linked to historical and sociological factors. During the middle of the nineteenth century the Protestant establishment in New England tended to characterize the Irish as a distinct race. At that time the Irish were said to display a range of primarily negative biological characteristics. Later, as the Irish gained in population and political power, this attitude changed.

One of the favorite and eternal arguments of dominant groups is that they *merit* their place in the social hierarchy (see also chapter eight). In times of absolutist royal power kings and nobles ruled through the doctrine of hereditary power. Ever since the enlightenment and the rise of industrial capitalism in the West, large segments of the middle class have rested their claim to social and political dominance on "social selection," a process said to be akin to natural selection. People might rise to the top from humble social origins, but if they did so it was on the basis of *merit.* According to this theory, merit was linked directly to heredity. Since the beginning of the twentieth century merit has come to be objectified as intelligence *plus* socially acceptable hard work.

Meanwhile, the somewhat vague concept of intelligence was converted into a supposedly measurable entity through the statistical concept of *IQ*. It is common for people to believe not only that IQ is hereditary but also that whole "racial" groups differ in average IQ. This has led to what is known as the "IQ argument." As noted above, the IQ argument was originally associated with class relations, particularly in Europe, and with race in the United States. In discussing the assumed link between race and IQ it must be made clear that we are about to deal with two nebulous concepts. We have already seen that race has no firm reality in biology. Now we need to examine the pitfalls in the concept of IQ as well as the notion that whole groups have different *hereditary* averages for intelligence.

What is intelligence? It should be obvious that tests designed to measure it are structured in relation to some theory, but I wish to delay discussion of this problem for the moment. Let us begin with test results and work backward to their origin and the concepts they reflect. Suffice it to say here that intelligence is a comparative phenomenon. Tests are standardized on the basis of a mean average in a population of test takers. Once this has been established individuals can be ranked above, at, or below the mean. It is also possible to give IQ tests to different groups of people, compare average scores, and rank one group against another. When such rankings are made in this country the data show that some sociological categories consistently score lower than the standard white American rage. These groups include American Indians, African Americans, and other ethic groups, such as Latinos of various origin. Class breakdown of scores shows that middle- and upper-class whites score better than lower-class whites and that people in the North score higher than those in the South. Much has been made of the fact that African Americans from the *North* scored better on one type of test (the army alpha given around the time of World War I), than *Southern* whites. When compared to *Northern* whites, however, they scored below the mean. The lower scores for African Americans in the North when compared to northern whites have been attributed to differences in social environment and education. It is important to note, however, that these results were couched in terms of differences among biological groups. But data actually concern four distinct *sociological* categories. These are: Southern *sociologically* defined whites, Southern *sociologically* defined

African Americans, Northern *sociologically* defined whites, and Northern *sociologically* defined African Americans. In no case do any of these groups represent a distinct biological population, although it *is* fair to say that the gene pools of each group differs from the others to some unquantifiable degree.

It must be stressed as well that IQ data is subject to variation in two ways. First, they depend on particular test protocols (many different kinds of IQ tests exist), and second, tests are given under varying conditions. It has been shown that *different* IQ tests produce *different* results and the *same* tests can produce *different* results when the testing conditions are varied.

Let us return to the concept of intelligence for a moment. What is it? The French psychologist Alfred Binet and his colleague, Theodore Simon, developed a set of tests between 1905 and 1911 that were meant to predict success in French middle-class elementary schools. They suggested that in intelligence there is a fundamental mental facility, the alteration or the lack of which is of utmost importance for practical life. That facility is judgment, or good sense, initiative, the faculty of adapting one's self to circumstances, judging well, comprehending well, and reasoning well. The concept of intelligence became a major preoccupation of American educational psychologists in the first half of the twentieth century. This led to various modifications of the definition. For example, Spearman (1904) reduced it to the ability to deduce relations and correlations. Thorndike (1927) regarded it as the power to make good responses from the standpoint of truth and fact. Terman (1937) defined it as the ability to think in abstract terms. All three definitions imply that intelligence can be measured as the rapidity of accommodation or adaptation to unique environmental situations through learning and conceptualization.

As the attention of psychologists turned to the concept of intelligence and how to test it, a debate began to emerge over the degree to which it is genetically determined. None of the definitions imply directly that IQ is genetic nor does anyone concerned with it claim that it is 100 percent hereditary. Rather, as I have already noted, arguments center over the relative contributions of heredity and environment to individual and group intelligence and how to measure objectively the contribution of each. Not so curiously, given the history of discrimination in the United States, most scholars who take a strong hereditarian position on IQ also assume that group differences in measured IQ are hereditary and that whites are genetically superior in intelligence to blacks. These arguments comfort the "racial" situation in the United States and have strong backing from many politicians and, unfortunately, psychologists (see epilogue), who have a strong academic stake in testing. The "new" field of evolutionary psychology (another name for sociobiology) reinforces the simplistic notion that most of human behavior is genetic in origin and that differences among cultural groups are biological in nature (see Wilson 1975).

Here it should be clear that we are concerned with two different problems. It is *not* the same thing to say that (1) genetics are responsible for individual differences in IQ within populations and (2) that population differences in IQ are due to genetics. The human species is highly *polymorphic.* Thus all human populations display wide internal variation in genetic traits, not to mention cultural variation. I have already noted that genetic differences *within* populations are wider than genetic differences among different populations. It is also true that, in the case of the human species, it is difficult, if not impossible, to separate genetic from environmental factors in the expression of phenotypic behavioral traits. Experiments designed to do this are impossible for both ethical and cultural reasons. We cannot breed humans in the laboratory the way we can breed rats. As we shall see in this chapter as well as the chapter on *The Bell Curve,* so far all attempts to empirically separate environmental from genetic factors in behavioral testing have been seriously flawed.

There are other problems as well. The selection and maintenance of a definite goal and the ability to criticize one's own behavior contain elements that are surely subject to environmental modification. No two children ever grow up under the same conditions. In addition, individual psychological differences other than intelligence may affect the way an individual responds to new situations. A hesitant, demurring child might do less well on tests than a more confident one even though both might have the same potential capacities. And, of course, when we deal with test results the testing situation must be considered, for these conditions are bound to be influenced by an individual's cultural and psychological background. Although intelligence tests are supposed to be self-contained units, that is, units that contain all the information necessary to make judgments within the context of the test, the intellectual background, interests, and experience of the individuals tested appear to have significant effects on the responses of test takers. In addition different cultures have a tendency to treat "truth" or "fact" in very different ways. In American society Aristotelian logic is imposed early on children, while in the East not only is a paradoxical type of logic taught in some areas, but some intellectuals there strive to change their basic thought patterns so that they can come to accept a kind of inversion of what Westerners would call truth: The obvious is always false; truth often lies in opposites.

IQ tests themselves are subject to artifactual errors that render interpretation difficult. Among these are such cultural factors as attitudes toward testing in general, the amount of test sophistication an individual brings to a particular experiment, and the structure of the tests themselves. Thus the tests may or may not actually measure what the experimenter assumes they do. For example, such tests are poor measures of biologically based *group* differences. The *independent* variable (the group being tested) is most frequently more *social* than biological. There are major problems with the *dependent* variable (what is being tested) as well. In addition to IQ, whatever that is, such items as motivation to take the test, intellectual background based on prior learning, and many other psychological factors affect test results.

Let me summarize those factors that go into an intelligent (successful) response to environmental stimuli. First, the nature of the stimulus must be considered. While there is evidence that the ability to respond to specific cues is partially inherent (humans in general have good vision and hearing and a relatively poor sense of smell), there is also evidence that a good deal of learning goes into the process. Perceptions are always selective (even when the process of selection is unconscious). The response to a cue involves such psychological factors as perceptual acuity, ability to discriminate among stimuli (which is partly hereditary and partly learned), and the ability to generalize, that is, to form classes of data from a range of sense perceptions. The latter is also clearly a process involving both learning and heredity. Accurate responses involve memory and the ability to retrieve necessary bits of information to be employed in problem solving. Interest and span of attention, both of which are highly dependent upon social and psychological factors, speed of response, and effectiveness of feedback from behavior are all important variables. No single gene, of course, could underlie all of these (and other) psychological processes. In addition each variable, dependent as it may be on a hereditary base, would be subject to environmental modification in different ways. Divergent behavioral phenotypes could emerge from the same basic genotype. This would arise through environmental shaping in the same way that similar phenotypes are derived from different genotypes.

Let us return to those "bright" and "dull" rat strains developed by Tryon. Later experimentation demonstrated that the rats were reacting to specific tests. Environmental factors had a strong effect on the performance of these inbred (and thus genetically pure) strains. More specifically, in three out of five maze measures "dulls" were either equal in performance or better than the "brights." The "brights" were more food-driven, low in motivation to escape water, timid in open field situations, more purposive, and less destructive. "Dulls" were not highly food-driven, were better on average in motivation to escape water, and were fearful of mechanical apparatus features. Note how important these additional facts are to a full understanding of Tryon's results.

Intelligence tests are designed to measure a series of abilities: for example, spatial relations, reasoning, verbal fluency, and facility with numbers. But these are no more culture-free than the concepts behind them. For although it is possible to define intelligence *operationally* as the ability to achieve high scores on IQ tests, we must never forget that certain socially significant concepts lie behind the operational definition. The major concept relates IQ to academic performance under existing forms of education. Our system of education, however, is geared to middle-class success, not necessarily innate ability. Arthur Jensen (1969a) has pointed out that many of the psychological properties that contribute to response potential intercorrelate, even though specific tasks such as spatial relationships, verbal analogies, and numerical problem solving might bear no resemblance to one another. In this he follows Spearman, who separated out a factor ("g") that he believed accounted for "general intelligence." This conclusion led Spearman to define intelligence as the ability to deduce relations and correlates. Nonetheless, Jensen, himself unsatisfied with a unidimensional concept of intelligence, delineated two *genotypi-*

cally (genetically based) distinct basic processes that he called level one (associational ability) and level two (conceptual ability). Jensen related level one to the formation of associations between related stimuli, red with danger, for example, and level two to concept learning and problem solving, for him by far the most important factor in intelligence.

Another author, Rosalind Cohen (1969), identified two conceptual styles that she called "relational" and "analytic":

> The analytic cognitive style is characterized by a formal or analytic mode of abstracting salient information from a stimulus or situation and by a stimulus-centered orientation to reality and is parts-specific (i.e., parts or attributes of a given stimulus have meaning in themselves). The relational cognitive style, on the other hand, requires a descriptive mode of abstraction and is self-centered in its orientation to reality; only the global characteristics of a stimulus have meaning to its users, and these only in reference to some total context. (Cohen 1969, 829–30)

For what I hope are obvious reasons, the analytic style is clearly correlated with success in the academic context. While they are perhaps not identical, the analytic style is certainly close to what Jensen (1969a, 114) refers to as level two learning, or conceptual ability. Cohen, however, rejects a genetic hypothesis and substitutes one in which socialization and group structure constitute the independent variables in the formation of cognitive style. I would claim that Cohen's arguments are more convincing than Jensen's.

> Observation indicated that relational and analytic cognitive styles were intimately associated with shared-function and formal styles of group organization. . . . When individuals shifted from one kind of group structure to the other, their modes of group participation, their language styles, and their cognitive styles could be seen to shift appropriately to the extent that their expertise in using other approaches made flexibility possible. It appeared that certain kinds of cognitive styles may have developed by day-to-day participation in related kinds of social groups in which the appropriate language structure and methods of thinking about self, things, and ideas are necessary components of their related styles of group participation and that these approaches themselves may act to facilitate or impede their "carriers" ability to become involved in alternate kinds of groups. (Ibid., 831)

As long ago as the 1930s C. C. Brigham, who had been convinced that IQ differences between immigrant groups could be objectively tested, offered this strong renunciation of his own past theoretical bias. "This review has summarized some of the most recent test findings, which show that comparative studies of national and racial groups may not be made with existing tests, and which show, in particular, that one of the most pretentious of these comparative racial studies—the writer's own—was without foundation" (Brigham 1930, 165).

Such candor is rare even in science. Once Brigham had reversed himself on culture-free testing he was able offer the following analysis for differential responses to IQ tests. Note how his discussion parallels the analysis of behavioral responses of animals to test situations.

> The assumption is made that people taking the alpha test [a U.S. Army IQ test used during World War I] adopted two different attitudes or sets, viz., a "directions attitude"—an attitude of careful attention to the examiner's instructions without looking at the test questions while the directions were read; and "reading attitude"—partially or completely ignoring the examiner's instructions while studying the test questions during the time in which the examiner was reading. The adoption of the first attitude would tend to give the individual higher scores in test 1 (entirely oral directions), test 6 (an unusual form of mathematical test), and 7 (a novel type of verbal test). On the other hand a person adopting the second attitude might quickly find out what was required in tests 3,4,5, and 8, and his score would be better if he ignored what the examiner was reading and studied the test questions during the period of instruction. (Brigham 1930, 162–63)

To my knowledge this analysis was the first published indication that the problem of constructing a "culture-free" test is not the only one in IQ testing. The procedures themselves appear to have strong and differential effects on the responses of individuals taking the tests.

As I have noted above, in the United States, for what are clearly social and historical reasons, the argument over IQ and heredity has centered around black-white differences. By 1966 Audrey Shuey could publish a heavy tome with the title *The Testing of Negro Intelligence.* Her summary of this issue ended with the conclusion not only that African Americans scored lower on most tests but also that the studies reviewed confirmed the hypothesis that differences between whites and blacks were largely due to heredity.

A search of the literature shows that this issue has been with us almost since IQ testing was invented. It was of course of great concern during the integration battles of the 1960s. Those who fought to maintain the status quo in the South argued that ending segregation would lower standards in the public schools through an influx of genetically inferior students. Those in favor of integration claimed that the poor showing of African Americans in the school system was an effect of segregation. This, they believed, was true in the North as well as in the South. Under then-president Lyndon Johnson in the middle 1960s, segregation as law was abolished officially in the entire country but continued de facto through residential patterns and educational inequality.

In an attempt to improve educational opportunity and to prepare young children for school, Project Head Start was begun in poor neighborhoods in the middle of the 1960s. Head Start nursery schools were designed to provide cultural stimulus for children before they entered kindergarten. While this project was welcomed by many in the country, there were those who felt that it was a waste of federal monies. Then, in 1969 a media bombshell struck. It was an article by Arthur Jensen, "How Much Can We Boost IQ and Scholastic Achievement?," published in what the press referred to as the "prestigious" *Harvard Educational Review.* By this time a new administration had taken over the White House. The country was in the midst of the Vietnam War, begun under Johnson, and priorities had shifted from domestic programs to foreign relations. Jensen's article, soon to be known as the "Jensen Report," argued that Head Start and programs like it were bound to fail. IQ was, he claimed, primarily hereditary, and African Americans were genetically inferior in IQ to whites. The Jensen article was reported in *Time* magazine to have shown up on the President's desk only a week after its publication. It was certainly taken seriously among those holding political power in the country.

I shall attempt to show below that in my opinion Jensen's paper fails the test of scientific validity. First, however, I should like to call attention to a point that has been overlooked by many on both sides of the IQ argument. Suppose, for a moment, that Jensen was completely right. He claimed that the total deficit between whites and blacks translated into 15 IQ points of difference and that of these, seven or seven and one-half points could be attributed to a genetic deficit among blacks. Now what could this seven to seven and one-half points mean in terms of educability? It should be clear that the answer must be: "Not much and probably nothing!"

Now to the failures of the report itself: Technically Jensen committed a major error when he claimed that his data concerning the degree to which IQ was hereditary, drawn as it was from white populations, could be used to speak about black populations. The genetic factor in IQ, estimated at 80 percent, was taken largely from studies by a well-known British psychologist, Cyril Burt. Burt based his conclusions on data from identical twins reared apart. Such twins are rare and important finds. They *are* identical genetically. Because they are raised in different environments, what measurable differences in behavioral traits occur between each individual in a set of twins can be attributed to the effects of the environment. If the difference between twins is 20 percent on a behavioral measure, for example, then one can say that the maximum effect of the environment is .20 and that the trait is, therefore, 80 percent genetically determined. The problem is that Burt's studies, even if they were correct (see below), were based exclusively on white twins. Additionally, one cannot say that the environments of identical twins reared apart are different enough to reveal genetic similarities as opposed to environmental ones. It is the general practice of adoption agencies to place children in homes that are similar in many respects to the homes of birth parents, except in cases where the birth parents maintained dysfunctional homes. It is even more likely that adoption agencies would go out of their way to place twins to be reared apart in similar homes regardless of the home situation of the birth parents.

The major statistic used in the Jensen Report is known technically as the *heritability* of a trait. The term can be very misleading. In fact, in biology it is employed in two contradictory ways. First, it is sometimes used loosely to indicate that a trait is genetic in origin. In this usage, when a trait is said to be "heritable" it simply means that it is in some way genetic. In the second, more correct and technical usage, heritability is *exclusively* a measure of *variance.* This means that it applies only when *variation* of some kind is present in a population. For example, the condition is satisfied when a population is made up of both blue-eyed and brown-eyed individuals. One can then ask the question: "What percentage of the variance (between blue and brown eyes) is due to heredity and what percentage is due to the environment?" The answer in this case is that the observed variation is 100 percent hereditary. Now if we are faced with a situation in which *everyone* is blue eyed, *then* there is *no* variance in the population (every individual is the same in reference to the trait in question, all have blue eyes). Because in this second case the trait does *not vary* among individuals the heritability is *zero* even though the trait is *100* percent genetic! This is a crucial point since laypeople are often confused by the term "heritability," and it is also misused even by some professional psychologists.

Because the concept of heritability deals with two variables, genetic *and* environmental, as a *statistic* it is subject to a very important restriction. No two populations ever live in exactly the same environment. If a trait with a genetic component is subject to environmental effects, as most are, these effects may differ in value from one environment to another. In other words the *penetrance* of the gene can be different in different environments. The genetic factor in height in humans is certainly based in large measure on heredity. But average height between two different populations might differ for such an environmental reason as nutrition (the degree to which proteins are found in the average diet of each population, for example). For this reason a measure of heritability in one population, let us say .80, is no guarantee that the heritability will be the same (for the same trait) in another population in another environment. It is always possible that the environment has acted differently on the same genetic potential. Thus the problem is that all the heritability figures available to Jensen came from white populations. What this means is that he had no right (from an experimental point of view) to extrapolate this figure for African Americans. Even in the most integrated parts of American society it is not possible to say that the environment for African Americans is identical to that of whites. IN FACT it is most likely that there are significant environmental differences for the two groups. Therefore, we have no idea what the heritability of IQ might be among African Americans. As the population geneticist James Crow put it in a response to Jensen published in 1969:

> It can be argued that being white and being black in our society changes one or more aspects of the environment so importantly as to account for the difference [in IQ]. For example, the argument that the American Indians score higher than Negroes on IQ Tests—despite being lower on certain socioeconomic Scales—can and will be dismissed on the same grounds: some environmental variable associated with being black is not included in the environmental ratio. (Crow 1969, 308)

Did Jensen know any of this when he wrote his report? Yes, he did. In an article published one year before his report he said the following:

> The inventors and developers of intelligence tests—such men as Galton, Binet, Spearman, Burt, Thorndike and Terman—clearly intended that their tests assess as clearly as possible the individual's innate brightness or mental capacity. If this is what a test attempts to do, then clearly the appropriate criterion for judging the test's "fairness" is the *heritability* of the test scores in the population in which the test is used. The quite high value of *H* for tests such as the Stanford-Binet attests to the success of the test-maker's aim to measure innate ability. . . . However, I would be hesitant to generalize this statement beyond the Caucasian population of the United States and Great Britain, since nearly all the major heritability studies have been performed in these populations. At present there are no really adequate data on the heritability of intelligence tests in the American Negro population. (Jensen 1968, 94)

The problem does not end here. Jensen's major heritability estimates were drawn from data provided by Cyril Burt (born 1883, died 1971). In 1972 and 1973 a Princeton University professor, Leon Kamin, began to speak out concerning what he saw as problems in Burt's data. Scientific models and the experimental data to support them rarely, if ever, show absolute statistical uniformity. Kamin became suspicious of Burt's material on heritability and IQ because it was just too good to be true. Burt published several studies of twins both reared apart and reared together. The correlation between IQ scores of the twins reared *apart* was given as 0.771. In addition Burt used a single statistic, 0.94, for twins reared *together,* again in every study. Kamin's criticisms were aired verbally in 1972. While Kamin had accused Burt of fraud, in a 1978 article published in

American Psychologist Jensen attempted to excuse Burt by saying that the peculiarities in Burt's data were probably due to carelessness. Jensen also claimed that whatever the reasons for Burt's data, they were no longer necessary to support his own arguments.

Kamin is not the only one to believe that Burt intentionally skewed the data to support his hypothesis. L. S. Hearnshaw (1979), an avowed fan of Burt's who gave the memorial address at the University of Liverpool on the occasion of Burt's death and who was chosen by the Burt family to write Burt's definitive biography, admits in that book that the evidence points to fraud, at least in the case of IQ: "The verdict must be, therefore, that at any rate in three instances, beyond reasonable doubt, Burt was guilty of deception. He falsified the early history of factor analysis . . . ; he produced spurious data on MZ twins; and he fabricated figures on declining levels of scholastic achievement. Moreover, other material on kinship correlations is distinctly suspect" (Hearnshaw 1979, 259).

A stronger argument, published in the *British Journal of Psychology* in 1983, was made against Burt by James Hartley and Donald Rooum. In a survey of Burt's work in the field of typographical research (one less likely to be controversial than heredity and IQ) they concluded:

> Sir Cyril Burt contributed to five main areas of typographical research: spacing words and text; the use of serifs; the effects of typefaces, type sizes and line-lengths on reading comprehension; and aesthetic preferences. Hearnshaw (1979) assessed this contribution as worthy of "well merited acclaim." In this article we examine what Burt had to say on each of these issues, and how far what is said is applicable to typographic practice today. It appears, despite the wisdom of some of the sentiments expressed, that many of them were opinions that were not supported by the data that Burt presented. Indeed there is possible evidence of deceit. We conclude, therefore (and Hearnshaw accepts) that Burt's contribution to typographic practice was marred by the same defects that one can find in his other work. (Hartley and Rooum 1983, 203)

Michael McAskie also disagrees with Jensen in reference to Burt. In a May 1978 article published in *American Psychologist* he argued that Burt's data "points more to invention than to genuine derivation." McAskie concluded by saying:

> It is a great pity that Jensen chose to write so ill-prepared a reply to the fraud allegations concerning Burt. Jensen does not even appear to have applied some of the tools of his trade in trying to distinguish between fraud and carelessness. He had no right to suppose that people suggesting fraud were merely speculating, nor was he particularly informed about the background of the *Sunday Times* article by Oliver Gillie or the political persuasions of those involved. "Sheer surmise and conjecture, and perhaps wishful thinking" are words that Jensen was not in a strong position to throw accusingly at others on this issue. (McAskie 1978, 498)

Perhaps one of the problems here is that Jensen was a post-doctoral student of Hans Eysenck (see chapter nine) who himself was a student of Burt's. What we have here is a (nongenetic) family connection. Beyond the fact that Jensen's work was based on Burt's statistics, Jensen's defense may, at least in part, be due to family loyalty.

It is apparent that Jensen accepts race as a valid biological division. Yet when Jensen talks about African Americans, the genes he is talking about (or better, a good percentage of them) come from a huge and varied continent. Thus, some analysis of ethnic and genetic diversity in Africa must be germane to the discussion.

Irving Gottesman (1968) in a book edited by *Jensen* (!) and others, discussed the geographic range of populations in Africa from which slaves were imported to Charleston during the period 1733 to 1807. His figures, taken from a study by William Pollitzer, show the following percentages: Senegambia 20 percent, Winward Coast 23 percent, Gold Coast 13 percent, Whydah-Bennin-Calibar 4 percent, and Angola 23 percent. Such a distribution covers more than a thousand miles of coastline and a territory extending for six hundred miles inland. The range of genetic and ethnic groups tapped was extensive.

In the United States itself, it is a vast simplification to speak of a single black or white genetic population. According to Gottesman: "The variation observed in the studies reviewed . . . are probably valid and reflect the genetic heterogeneity of Negro Americans living in different geographical and social distances away from their white neighbors. Such heterogeneity prevents us from speaking validly of an average Negro American with x percentage of white genes" (Gottesman 1968, 20).

In sum, genetic studies of black versus white intelligence (whatever that is) based upon undifferentiated U.S. samples are naive in the extreme because they do not consider distributions of genetic variation in either Africa or the United States.

The problem does not end here. As we have seen, Jensen found an overall intelligence deficit of 15 percentage points among African Americans. He was willing to attribute about half of this difference to environmental influences. The other 7.5 points were then assumed in the report to be due to genetic factors. Yet on page 100 of his 1969a article Jensen states:

> In addition to these factors, something else operates to boost scores five to ten points from first to second test, provided the first test is really the first. When I worked in a psychological clinic, I had to give individual intelligence tests to a variety of children, a good many of whom came from an impoverished background. Usually I felt these children were really brighter than their IQ would indicate. They often appeared inhibited in their responsiveness in the testing situation on their first visit to my office, and when this was the case I usually had them come in on two to four different days for half-hour sessions with me in a "play therapy" room, in which we did nothing more than get acquainted by playing ball, using finger paints, drawing on the blackboard, making things out of clay, and so forth. As soon as the child seemed to be completely at home in this setting, I would retest him on a parallel form of the Stanford-Binet, a boost in IQ of 8 to 10 points or so was the rule; it rarely failed, but neither was the gain very often much above this.

Was Jensen unaware that these are the conditions that are not met by the majority of studies he cites in his report, particularly those drawn together by Shuey (1966)? If the deficit he notes is consistent in disadvantaged children, then all the IQ differences noted between whites and blacks in the United States may be subsumed under a combination of testing errors and environmental effects.

The Jensen Report contains other distortions and misinformation concerning cited data. The following material was extracted by Dr. Carol Vance and myself from a close reading of the Jensen Report and a comparison of his citations with what was actually said in the original sources.

On page 23 of the report, Jensen refers to an article by Cyril Burt (1963). He says that in the general Negro population there is an excess of IQs in the 70–90 range (see Jensen's illustration on page 25 of the Report). This excess is explained as the combined effects of severe environmental disadvantage and emotional disturbance, both of which act to depress test scores. On page 27 Jensen says that Burt corrected for this bulge by eliminating scores of those having depressing factors. However, according to the original Burt article there is a lack rather than an excess in 70–90 range (see figure 1 in Burt 1963, 180).

On page 40–41, Jensen cites Cooper and Zubek (1958). He stresses the effects of rearing bright rats in normal and enriched environments and says, "While the strains differ greatly when reared under 'normal' conditions . . . they do not differ in the least when reared in a 'restricted' environment and only slightly in a 'stimulating environment.'"

Our reading of the same article puts things the other way around. Cooper and Zubek stress the benefits of stimulation to dull animals. "A period of early enriched experiences produces little or no improvement in the learning of the bright animals, whereas dull animals are so benefited by it that they become *equal* to bright animals. On the other hand dull animals raised in a restricted environment suffer no deleterious effects, while bright animals *are* retarded to the level of the dulls in learning ability" (Cooper and Zubek 1958, 162). This result extrapolated to humans supports the hypothesis that deprived environments such as those known to exist for the poor and particularly black Americans should have an effect on IQ scores.

If one compares Jensen's figure 6 on page 50 of the Report with figure 1 of Erlenmeyer-Kimling and Jarvik's (1963) article, from which some of Jensen's data is drawn, we find that Jensen shows only the midpoints for correlations between relatives reared together and reared apart. This emphasis stresses the discreteness and difference among the correlational scores while the original diagram, which shows the range and the median, demonstrates the overlap of correlational range and hence an overlap in the strength of genetic inheritance.

On page 63 Jensen cites a study by Wheeler (1942) of IQ among Tennessee mountain children and notes that environmental improvements do not counteract a decline in IQ of "certain below average groups." Jensen neglects to mention Wheeler's discovery that the decline in IQ is due to the large percentage of held-back children. This factor raised the age level in every grade and therefore depresses IQ scores because these are always correlated with age. When Wheeler separated out the scores of older children in each grade he found that the other children performed normally. Comparing chronologically "true" members of each grade over time (with those overage weeded out) he found that in most years there was no decline. Wheeler says that the chronological IQ drop of 20 points is accounted for by children being repeatedly held back, which means more older children will be found as the grades get higher. Their presence depresses IQ scores most in the higher grades. If Wheeler's logic is followed the decline that Jensen presents as ranging from 103 to 80 points of IQ is reduced to the range 102.76 to 101.00 points!

On page 74 Jensen says that on the average first-born children are superior mentally and *physically* to their siblings. His citation here is Altus (1966). Altus, however, presents no evidence about physical superiority. Altus does cite a study by Huntington showing differences in birth order and achievement that suggests that the differences are caused by superior physical strength of the first born. Altus has the following to say about Huntington's hypothesis: "While his finding is typical of all those reported thus far, his explanation of the linkage is *not* typical: He argued that the first born probably tend to be physically stronger and healthier. . . . *One may safely accept his data on the birth order of the eminent without accepting his explanation*" (Altus 1966, 45; italics mine).

On page 76 Jensen cites Burt's (1961) contention that the inheritance of intelligence conforms to a Mendelian, polygenic model. Yet, he fails to note the wide variety of intelligence within a social class and the fact that children's scores are not as narrow as those of their parents. In fact, if there were no social mobility at all and class were totally static, the result of breeding over five generations would be a disappearance of class means. "After about five generations the differences between the class-means would virtually vanish, and the proportional range within each class would spread out almost as widely as the proportional range of the population as a whole" (Burt 1961, 15).

Other British studies show that IQ scores within social classes have been remarkably stable over the past hundred years. This is because bright lower-class children have moved up the social scale while less bright upper-class children have moved down. Burt's study appears to bear this out for England. Needless to say the notions of bright and less bright used here do not necessarily imply genetic differences, although they might.

Now if the same model is applied to African Americans, intelligence would have remained constant by class if social mobility operated as it is supposed to operate in England. But even in the lowest class, there would be children of above-average intelligence who would rise so that the range of child intelligence would be much wider than adult intelligence. This is the process known to statisticians as "regression to the mean." In any case, Jensen does not mention this aspect of African American performance, that is, unexpected *over* performance.

In any case the model cannot be applied in the United States, because when Jensen published his report little real social mobility existed for African Americans. Even today, in spite of some increased social mobility, African Americans do not experience the same degree of social mobility as whites. Additionally, it is necessary to stress yet again that from the point of view of genetics, blacks in America do not constitute a homogeneous population.

On page 83 Jensen cites research by Heber and Dever on education and habilitation of the mentally retarded. While we did not have access to Jensen's original source for this citation (a paper read at the Conference on Sociocultural Aspects of Mental Retardation), we did read a paper by the same authors entitled "Research on Education and Habilitation of the Mentally Retarded." It appeared in *Social-Cultural Aspects of Mental Retardation,* edited by H. C. Haywood (1968).

Jensen says that Heber has estimated that IQs below 75 have a much higher incidence among African American children than among white children at every level of socioeconomic status (Jensen 1969a, 83). We found no statement by the cited authors that African Americans have a higher frequency of IQs under 75 than whites. Furthermore, Heber's study was not a study of race and intelligence but rather a study of a special group of mentally retarded children from a specific neighborhood in Milwaukee that was:

> Characterized by having the city's highest known prevalence of mental retardation among school age children. The nine census tracts which compose this area, known as the "inner core," also have the city's highest rate of dilapidated housing, the greatest population density per living unit, the lowest median income level, and the greatest rate of unemployment. Though comprising no more than five percent of Milwaukee's population it yields about one-third of the mentally retarded known to the schools. (Heber et al. 1968, 35)

While it is a good bet that this population is composed primarily of African Americans given its socioeconomic profile, the point of Heber's study was to show that much of what passes for mental retardation is caused by *cultural* rather than genetic factors. One might also wish to take into account the degree to which slum dwellers in urban America are exposed to a high percentage of lead poisoning due to the ingestion of lead dust in old, poorly cared for housing. It is a well-known fact that lead poisoning has a strong effect on mental capacity, particularly in children.

On page 86 Jensen cites a study by Geber (1958) that discusses precocity of African American infants. Jensen mentions motor precocity but neglects to mention intellectual development as well. Geber says, "The result of tests showed an all round advance of development over European standards which was greater the younger the child. The precocity was not only in motor development; *it was found in intellectual development also*" (Geber 1958, 186).

The main thrust of Jensen's paper, which has been somewhat buried by popular accounts, is that there is a wide diversity of mental abilities in humans and that educational programs should be tailored to meet the needs of all children. It is difficult to disagree. It is most unfortunate, however, that Jensen pleads this case in the context of a report centered on a flawed discussion of genetics and IQ. In his report Jensen took a fairly safe, if as yet unproved hypotheses—that intelligence is heritable (that it varies among individuals by genetics and environment)—and forced it to carry the burden of a second argument for which there is no acceptable evidence at all.

In 1977, Jensen published an article in *Developmental Psychology* ("Cumulative Deficit in IQ of Blacks in the Rural South"). Here evidence *is* offered in support of an environmental explanation for IQ deficit! In this study Jensen finds substantial decrements in IQ as a linear function of age and relates it to educational differences. This study did not lead Jensen to change his mind, however. Instead he compares his new data with a previous study of children in Berkeley, California. (In the Berkeley study Jensen found no significant decrements in IQ in either his white or black sample.) This led him to conclude:

> However, the present results on Georgia blacks, when viewed in connection with the contrasting results for California blacks, would seem to favor an environmental interpretation of the progressive IQ decrement [in Georgia]. If the progressive IQ decrement were a genetic racial effect per se, it should have shown up in the California blacks as well as in the Georgia blacks, even if one granted that the California blacks have a somewhat larger admixture of Caucasian ancestry than do blacks in Georgia. . . . But the California blacks showed a slight, though significant decrement only in verbal IQ, which one might expect to be more susceptible to environmental or cultural effects than nonverbal IQ. The blacks of rural Georgia, whose environmental disadvantages are markedly greater than in the California sample, show considerable decrements in both verbal and nonverbal IQ. (Jensen 1977, 190)

Apparently Jensen refuses to consider the possibility that even in Berkeley, California, the social and educational environments for whites and blacks might be different and have an effect upon IQ test scores. It might be useful, therefore, before closing this chapter to look at some of the sociological situations that affect the performance of children on IQ tests in American society.

Two studies dating from the 1960s (Cohen's work cited above is also relevant) have amplified the role of culture and social group in both test results and academic performance. Katz (1968) varied test conditions for samples of African American students in relation to "subjective probability of success"—that is, how the individual taking the test feels about how he or she will score. Differences in this attitude were then measured against different types of testing situations in which the race of the tester was varied as well as the kinds of attitude expressed during the testing situation. The theoretical basis for this study comes from the psychological concept of "need achievement" in which "the strength of the impulse to strive for success on a given task is regarded as a joint function of the person's motives to achieve, the subjective problems of success, and the incentive value of success. According to the model, on a test that has evaluative significance (e.g., a classroom test) motivation is maximal when the probability of success is at the .50 level."

Katz notes that in a number of experiments with black college students, individuals tend to underperform on intellectual tasks in the presence of whites. Katz speculates:

> . . . that for Negroes who find themselves in predominantly white academic achievement situations, the incentive value of success is high but the expectance of success is low because white standards of achievement are perceived as higher than own-race standards. By the same token, the perceived value of favorable evaluation by a white adult authority is high, but the expectancy of receiving it is low. Therefore, by experimentally controlling Negro subjects' expectancy of success on cognitive tasks it should be possible to produce the same, if not higher, levels of performance in white situations as in all-Negro situations. (Katz 1968, 134)

A group of freshmen were given a test that was described to them as part of a scholastic aptitude test. They were told that their scores would be evaluated in comparison to scores achieved in predominantly white colleges. The students were given a pretest and then told what their chances of success on the actual test would be. One-third were led to believe that they had little chance of meeting the standards for their group, one-third were told that they had an even chance, and one-third were told that they had a good chance. Each of these three groups was then divided into subunits, one given a white tester, the other a black tester. "The results showed that in the low and intermediate probability conditions, performance . . . was better with a Negro tester, but when the stated probability of achieving the white norm was high, the performance gap between the two tester groups closed" (Katz 1968, 134).

Another test, in which a college with no admission standards other than high school graduation was compared to a college with high relative standards, demonstrated that the effects of varying the race of the tester were the same as in the controlled experiment described above. On the other hand, the scores achieved by students at the selective college were higher when the testers were white, no matter what the probability of success. Katz explained these differences:

> In summary, it appears that Negro students who had been average achievers in high school (the non selective college sample) were discouraged at the prospect of being evaluated by a white person, except when they were made to believe that their chances of success were good. But Negro students with a history of high academic standards (the selective college sample) seemed to be stimulated by the challenge of white evaluation, regardless of the objective probability of success. (Katz 1968, 138)

Katz generalized his results in terms of differences in socialization between lower- and middle-class children. "The present assumption is that lower class children . . . because they have received less parental approval for early intellectual efforts remain more dependent than middle class children on social reinforcement when performing academic tasks" (Katz 1968, 138).

While Katz' s experiments do not relate directly to intelligence testing they do go a long way toward explaining why certain sociological groups respond as they do to education. The problem is complicated since it involves the motivation of the individual, which is partly a product of home experience but also of the students' conception of the expectations of teachers defined partially in terms of race. The common educational experience of lower-class African Americans with white *and* black teachers is often discouraging. This problem is amplified by the environmental setting in which the probability of success is lowered by the experience of daily life. All these factors would act to lower success in any testing situation.

The process of learning in children is even more subtle than Katz's findings would indicate. A study of performance of children in the San Francisco schools supports the hypothesis that a teacher's attitude toward the success of a child will have a profound effect on the outcome of the educational process.

The experimenters established the expectation in teachers that certain children in the school chosen at *random* would show *superior* performance in the coming school year. This expectation was established by testing the children on an intelligence test and *mis*informing teachers of the results. The use of this test in the pre-experimental situation had the added advantage of providing a true measure since the children could be reexamined with the same test later in the experiment. A casual method of informing the teachers about the presence of "potential spurters" was used. "The subject was brought up at the end of the first staff meeting with the remark 'By the way, in case you're interested in who did what in those tests they were done for Harvard'" (Rosenthal and Jacobson 1968, 22).

All the children were retested four months after school started, at the end of the school year, and finally in May of the following year. As the children matured, they were given tests appropriate to their level. These were designed to evaluate both verbal skills and reasoning. The results showed that children who were expected to do well by teachers showed the greatest intellectual gains. An unanticipated finding of the study was that when teachers were asked to evaluate the undesignated children, many of whom had gained in IQ during the year, they tended to evaluate them negatively. The more they gained the less favorably they were seen!

Since writing his report Arthur Jensen has continued his work in the field of IQ and race. It seems to me that in these works Jensen ignores all the recent evidence for the nonexistence of racial categories. Among his recent works are Jensen 1974, 1977, 1978a, 1978b, 1980, 1984a, 1984b, 1985, 1986, 1987a, 1987b, 1989, 1990, 1992, 1993a, 1993b. In 1999 he wrote a laudatory comment in the pretext pages of J. Philippe Rushton's book on race, IQ, and brain size (criticized in chapter nine of this book). In none of his works that I have seen does Jensen take account of the evidence against the existence of race as a valid category in the classification of humans. Nor does he seriously discuss the many studies that support the flexibility of IQ in the context of environmental differences. Instead he continues to argue as he has always done, relating race to IQ with the persistent claim that the average black is inferior in intelligence to the average white. To my knowledge he has never cleared up the contradiction between his understanding of the concept of heritability, which, as noted above, he admits does not allow for intergroup comparisons, and his consistent use of heritability statistics for samples of whites applied to blacks.

In the next chapter I will discuss the IQ argument from the perspective of Konrad Lorenz and Robert Ardrey. Lorenz was a Nobel Prize winner and internationally acclaimed biologist, and, although his research concerned fish and birds, he ventured well beyond his competence to speculate, in a series of popular books and articles, about human behavior. Robert Ardrey authored three bestsellers, popularizing Lorenz's ideas concerning our species. Although Lorenz's and Ardrey wrote before the official founding of sociobiology by E. O. Wilson, their extreme form of biological determinism, based primarily on extrapolations from nonhuman animal behavior to humans, stands between the vulgar biological determinism, of the nineteenth century and the explosion of a somewhat more scientific form in the last quarter of the twentieth. Both Lorenz and Ardrey preached the doctrine of racial purity and both argued for the inferiority of certain races.

CHAPTER SIX

BIOLOGICAL DETERMINISM AND RACISM

Robert Ardrey and Konrad Lorenz

In this chapter I discuss the work of Konrad Lorenz, the father of ethology (the study of animals in their native habitat), and Robert Ardrey, a political conservative and extreme advocate of biological determinism. Although it was not a major element in either man's work, both published racist arguments that flowed from their ideas about how evolution worked. During the Hitler period Lorenz wrote an article in which he argued that race mixture was a menace to the physical and moral health of the German people. After World War II ended Lorenz declared himself a social democrat, but his old ideas about racial purity resurfaced in somewhat sanitized language in his book *The Eight Capital Sins of Our Civilization.* More about this book later. Let me say now, however, that in my experience and in the experience of several of my colleagues who also teach courses on race and racism, we find that students often refer positively to the racist ideas expressed by Ardrey and Lorenz in their books, which are still widely available in the large chain bookstores. I have also noted that beyond the classroom the names Lorenz and Ardrey are often cited to support many of the arguments argued against in this book.

Robert Ardrey successfully recycled himself as an advocate of Lorenz's ideas, particularly the role of aggressive behavior and territoriality in natural selection, including human evolution. He is the author of three best-sellers on the topic: *African Genesis, The Territorial Imperative,* and *The Social Contract.* In what follows I shall discuss the parallels between Lorenz's and Ardrey's thinking and how these lead to their convictions concerning notions of inferiority and the consequent necessity for racial purity.

First a disclaimer: I am not adverse to biology and see myself as a confirmed Darwinian. I agree that human beings are certainly not exempt from biological rules. At the very least we must all respond to the minimal requirements of our physiology in order to survive. These requirements are species wide, but the responses vary from society to society. In general they are cultural (learned) rather than genetically determined. However, it is also the case that humans share an evolutionary past with all living forms, from distant single-celled organisms to our closest relatives, the monkeys and apes. This heritage includes aspects of behavior that are programmed in our genes. All humans, for example, share the mammalian *capacity* for aggressive behavior. Whether or not this capacity is *expressed,* and how it is expressed, however, is mediated by culture.

The relationship between biology and human behavior may be more specific and stronger than the mere fact that we share biological capacities for certain types of behavior. All normal humans, for example, learn to speak a language, and while the differences among languages are historically and culturally determined there is good reason to believe that a universal grammar (a grammar of grammars) is coded in the brain. In the last thirty years this theory, first proposed by the linguist Noam Chomsky in the late 1950s, has been supported by a growing body of evidence. Chomsky's theory speaks to a trait shared by all humans. It makes no claim to explain different languages on the basis of heredity. Chomsky also supposes that language is a new trait exclusive to humans. Thus, he denies that apes can be taught language because they are not equipped with a genetic program for it.

While Chomsky's theory attempts to account for an exclusively human behavioral pattern, other biological determinists have focused on what they believe are evolutionary continuities among all animal species.

Sociobiology, which attempts to explain altruistic behavior, is such a theory. I am not a fan of sociobiology as it applies to humans, but one thing can be said for it—it is not a priori racist, nor are all other biological determinists a priori racist. Such theories become such only when they attempt to explain behavioral differences among human cultural groups on the basis of genetics. Such is the case for Konrad Lorenz and Robert Ardrey, the latter the only nonacademic to be discussed in this book.

The counterattack on the assumed dictum of twentieth-century social science—that the proper study of mankind is of man alone—was led in the professional world of biologists by Konrad Lorenz. While Lorenz attempted to draw a synthesis between the biological heritage in human behavior and its expression in culture, Robert Ardrey hewed to a much stricter biological model. According to him, if we understand human biology we understand human behavior in all its facets. For both authors the main focus was human aggression, which Ardrey believed operates to insure an even more fundamental instinct—territoriality. In discussing this issue the old playwright in Ardrey comes screaming through. He puts the question so dramatically that a reader caught in his rhetoric may fail to spot its simple-mindedness. "How many men have you known in your lifetime, who died for their country: And how many for a female?" (1966, 6). The answer, of course, for Ardrey is more "for their country."

Melodrama is not Ardrey's only technique. His argument often begins with sound scientific data, only later slipping into a series of non sequiturs. To maintain his argument he makes it appear as if the experts are firmly in his camp. Ardrey invites the layperson to join in an alliance to wage a battle against a band of scientific nitwits who have been the agents of a conspiracy to rob the public of access to scientific facts. This attitude is reinforced by Ardrey's reliance upon a form of reverse *ad hominem* in which the reader is warned that in joining him they will be attacked by liberals, Marxists, and Freudians. We shall see later that this same technique has been used by many of the other individuals discussed in this book. This conspiratorial theory is called forth as we are told that social scientists attracted to the classless state (read: Marxists) have forgotten that "hierarchy is an institution of all social animals and the drive to dominate one's fellows an instinct three or four million years old" (1961, 13).

Ardrey claims that none of us can differ in significant measure from the earliest members of our species. No instinct, physiological or cultural, that was part of the original human bundle can ever be permanently abandoned. Besides the fact that the term *cultural instinct* is a contradiction in terms, this statement is complete nonsense for it reflects a dismal misunderstanding of genetics in which the penetrance of a particular gene is the complicated outcome of environmental and genetic processes. Genes are suppressed or modified by other genes in the genetic system and by the effects of the environment on the maturing system.

Reading Ardrey, it would seem as if science itself were driven by instinct. "The contemporary revolution in the natural sciences, unorganized, undirected, and largely unrecorded, has with a strong *instinct for survival* challenged the romantic fallacy in a voice unlikely to be heard" (1961, 16, italics mine). The enemy is clear. For Ardrey the naive social scientist has been the victim (when he is not the perpetrator) of the "romantic fallacy." This is the belief that humans are born good and that the evil emanating from human behavior is due to the effects of the social environment. Following the fallacy, all one need do to reach utopia is modify the environment and our basic goodness (our primitive human nature) will emerge. This is, to say the least, a peculiar analysis. At best, it provides a parody of Freud and Marx, as well as most of twentieth-century social science. Even the most behavioristically inclined psychologists, sociologists, and anthropologists have been anything but romantic about humans. They do not see our, or any other, species as good or evil, but rather as value neutral. For what can "good" or "bad" mean anyway, unless these terms are defined by humans in the context of socially derived ethics? In fact, most behaviorists have been overly cautious in avoiding value terms in theories, which are based on what they hope are objective observations. It is not an exaggeration to say that value terms give behaviorists the hives. To be sure neither Marxists nor Freudians are behaviorists. Marx, however, believed that human values change as they are shaped by history. Freud was deeply suspicious of any theories suggesting that humans were less than highly selfish and egotistical animals. One need only read *Civilization and Its Discontents* to see what a dark view Freud took of our species as well as of civilization in general.

Both Ardrey and Lorenz substitute one unproved doctrine for another, that of original sin! This puts them into the camp of Old Testament moralists, but with a biological twist. While Ardrey never expresses this overtly Lorenz makes an explicit statement: "All the great dangers threatening humanity with extinction are direct consequences of conceptual thought and verbal speech. They drove man out of the paradise in which he could follow his instincts with impunity and do or not do whatever he pleased" (1961, 29).

For Ardrey man is simply a "killer." To support this thesis he draws upon a dead argument borrowed from the speculations of the anatomist Raymond A. Dart, who was, to his credit, one of the first scientists to suggest that tools were a major force in human evolution. Dart claimed that our fossil ancestors, the *Australopithecines* of Africa, were the first primates to use tools as an aid in hunting. But both Dart and Ardrey went beyond this to suggest that such tools, some of which may have been weapons, were all artifacts of an aggressive species. Analysis of the crude chipped stones and bones found with some *Australopithecine* species suggests that they were used for a variety of purposes, among them hunting. It is important to note, however, that hunting and aggression cannot be equated. One involves the killing of other species, the other fighting and/or killing within one's own species.

At the time Ardrey wrote there was some substantive evidence and much theory to support the idea that the transition from ape to humans involved a change in diet from one based largely on vegetable products to one based largely on meat. These theories suffered from one major defect, however. They were based on models limited to what the males of the species were doing. At present we are beginning to understand that an omnivorous diet involving *collecting* as well as hunting, and cooperation between collectors (often women) and hunters (usually men), was of major importance in the emergence of our species.

But for Ardrey the "weapon" was the "father" of the first humans. Among modern physical anthropologists the late Sherwood Washburn and his students at the University of California at Berkeley have been in the forefront of those who have attempted to show a relationship between tools (not just weapons) and the rapid evolution of our species. The argument goes as follows: Tool use presented an immediate advantage for a rather weak species because it was of significant aid in the food quest. Any development in the direction of improved tool manufacture and use would produce a concomitant advantage for those involved. Tool use, however, requires a certain physical structure, the most important element of which is upright posture. Washburn suggested, therefore, that the first advance toward *Homo sapiens* was the development of true bipedalism. This freed the hand for carrying and tool use. The intelligent use of tools, however, requires special cerebral organization producing higher intelligence and good eye-hand coordination. Better tools in combination with a cooperative social organization favor individuals with better brains, and better brains produce more complex and better tools. Communication systems, and eventually language, develop as a side effect of this process. Language, of course, is the sine qua non of human culture. The rapid change in morphological structure of fossil hominids that began with upright posture and an anatomy adjusted to walking erect finishes with the development of more advanced brains. The use of tools for hunting and gathering increases the calorie-getting efficiency of the group and allows for an expanding population. As populations expand some groups migrate to new territories. Environmental differences between new and former territories stimulate further cultural adaptation.

In chapter two of *African Genesis,* Ardrey slips into his territorial thesis, which is developed in greater detail in his next book, *The Territorial Imperative.* Ardrey overstates his case at the outset. "When we find a characteristic prevalent among all branches of the vertebrates, such as the instinct to maintain and defend a territory, then we must mark it a significant instinct indeed" (1961, 71). "Every primate species so far studied with the significant exception of the Gorilla maintains and defends a territory" (ibid., 720). In fact among our species' closest relatives (the simian apes: gorilla, chimpanzee, orangutan, and gibbon) *only* the gibbon is clearly territorial and aggressive in its natural habitat. Allison Jolly, a zoologist who has published extensively on lemur behavior, has this to say about aggression in the primate order:

> The level of aggression and the means of controlling it seem to have suppressed the expression of aggression within a troop almost entirely—the howlers, the gorillas. In others—particularly baboons, and macaques—

> aggressive interactions may be common. *Aggression among the same species in different environments also varies immensely.* Baboons and vervets in lush Ugandan forests may have far fewer aggressive interactions than those of the Tanganyikan plains. (Jolly 1966, 154; italics mine)

Ardrey tells us that humans as predators have a natural instinct to kill with weapons. The question is, to kill *what* with a weapon? And how could this "instinct" have arisen? Could it have been that the first humans felt unquenchable urges to seek out first a weapon and then a victim? Is this the aggression that both Ardrey and Lorenz admit is a relatively passive phenomenon? Lorenz has an answer for this paradox when he suggests that humans' cultural means for killing has outstripped the biological restraints generally built into the aggressive response, but nowhere does he suggest that humans have an *instinct* for killing, much less killing with a weapon, which is after all a *cultural* object.

Lorenz's *On Aggression* is a somewhat more modest book than any of Ardrey's three tomes. Lorenz, unlike Ardrey, does not pretend to have all the answers to the puzzle of human existence. He is also more scientific in his approach. For Lorenz, who spent much of his life studying birds, aggression was the key for understanding social organization in animal groups. It functions to maintain social equilibrium. It is only among humans, because of culture (rightly seen by Lorenz as a nonhereditary form of adaptation), that aggression provides the basis for self and group destruction. Thus, for Lorenz the dialectic is one of conflict between the human species' biological drives, particularly the aggressive drive, with our exaggerated cultural means of displaying it, and human ethical systems that are, if properly developed, capable of containing it. Lorenz sees aggression in all animals, including humans, as a spontaneous instinct that *must* find gratification. "If stimuli to release it fail to appear for an appreciable period, the organism as a whole is thrown into a state of general unrest and begins to search actively for the missing stimulus" (1966, 50).

With one exception the only evidence for this theory is drawn from material on nonhuman animal behavior. Lorenz employs the disarming technique of anthropomorphizing the animals he so obviously loved, at the same time reducing humans to animals through zoomorphization. The result is most peculiar. Geese feel "love" and "jealousy." Scapegoat behavior in humans is seen as closely analogous to behavior in certain types of fish! Humans are trapped by their instincts, and culture becomes an impediment to development. Human beings would have been happier if they had remained in a state of nature, where instinct would have served their needs without the dangerous and distorting effects of tradition.

In spite of what Lorenz says, analogous aggressive responses in certain of the lower animals and humans differ in an important respect. In aggressive animals *specific releasing* stimuli elicit a stereotypic response. This is not the case among humans. Releasers, far from being inborn, are learned. This can be said absolutely, for such releasers are symbols defined in a personal and/or cultural way. For human beings the same stimuli may be either neutral, aggression stimulating, or have some other symbolic function depending on what an individual has learned. In contrast, it is the very predictability of response to specific stimuli that makes Lorenz's animal data so convincing. The difference between aggression in lower animals and humans is that in *some* lower animals it is instinctual and stereotyped while humans act aggressively (when they do) for a variety of cultural reasons. The only thing that can be said with certainty about the biological basis of aggression in humans is that it *is* an inborn *capacity* of our species.

The culmination of *On Aggression* hinges upon an unpublished study of the Ute Indians (by Sidney Margolin) discussed by Stewart (1968). This is the unique case cited to support an assumed instinct for aggression in humans. Furthermore, Margolin's material is used to prove that natural selection has operated to favor the development of an *extreme form* of the aggressive drive in one ethnic group. Additionally, Margolin (and Lorenz) claimed that the blockage of aggression among these Indians led to the highest frequency of neurosis in any human society. The cause of this neurosis was said by Lorenz to be "undischarged aggression" (1967, 236).

Where are the controls? How do we know that this ethnic group suffers more frequently than any other from neurosis, and if it does how can we say that the neurotic behavior is due to suppressed aggression? With how many other ethnic groups has Lorenz compared the Ute? Anthropologists familiar with them find no similarity between what Lorenz says and their own knowledge. Permit me to quote extensively from an answer to Lorenz and Margolin by an expert on the Ute Indians, Omar C. Stewart:

> Plains and prairie Indian ethno-history does not reveal that these people led a wild life consisting almost entirely of wars and raids. They had war and they did raid, but their history was not of such war and raiding so as to make them unique among social groups. The Ute Indians were early horsemen and hunted buffalo on the high plains. While there, they were prepared to fight or run, and they always returned to the Rocky Mountains and maintained their homes there. In early historic times they were often referred to as the "Swiss" Indians.
>
> The implication that these Ute were a violent people, addicted to war, is not borne out by historic facts. . . .
>
> Finally the peyote religion of the Ute is a significant refutation of the Lorenz-Margolin thesis. The Ute appear to have been receptive to peyote as early as 1900, according to Mooney, who reported their participation in the peyote ritual with the Jicarilla Apache. Their interests grew by means of visits to Taos and Oklahoma until the cult was firmly established, especially on the Mountain Ute reservation in southwestern Colorado where it flourished from 1916 on. The peyote religion is a syncretistic cult, incorporating ancient Indian and modern Christian elements. The Christian theology of charity, and forgiveness has long been added to the ancient Indian ritual and an original desire to acquire personal power through individual visions. Peyotism has taught a program of accommodation of over 50 years and the peyote religion has succeeded in giving Indians pride in their native culture while adjusting to the dominant civilization of the whites. (Stewart 1968, 105–8)

Another author, John Beatty, also takes issue with Lorenz concerning the Ute. He says the following:

> To be consistent with Margolin's scheme, one must assume that only those who were aggressive would be permitted to mate.
>
> These assumptions are not supportable. Insofar as inbreeding is concerned, we find that Ute marital patterns are not solely Ute or Ute prairie-tribe marriages. As a matter of fact, the Ute engaged in some warfare, basically raids, which were carried out primarily for the purpose of women stealing. These raids were made on non-Prairie Pueblos. The Pueblos have long been described as non-aggressive. . . .
>
> Rockwell states that in addition to out marriages with the Pueblo tribes the Ute also married the non-Prairie Bannock and Snake Indians living to

> the West. This would further invalidate the conclusion of Lorenz and Margolin, since it was stated that only prairie tribes "were subject to the selection pressures described." Since the Bannock and Snake or Pueblo Indians do not qualify for the "selection pressures," we must assume that they are not aggressive peoples and are, therefore, introducing genetically different material into the Ute gene pool. (Beatty 1968, 107–8)

This is a crucial argument, since Lorenz has attempted to provide a behavioral-genetic basis for understanding Ute social behavior. If the marriage patterns do not conform to the process Lorenz and Margolin describe, the entire hypothesis fails.

The inability of Lorenz to connect this hypothesis rigorously to sound genetic evidence is no more obvious than in his contention that *intra*specific selection is still operating among humans in an undesirable direction. He says that there is a high positive selection premium for instinctive traits such as the amassing of property, self-assertion, etc., and that there is an almost equally high *negative* premium on simple goodness. He warns that competition in the business world threatens to fix in us hypertrophies of these traits as horrible as the intraspecific aggression evolved by competition between warring tribes of stone-age man.

The reader may wonder why I have spent so much time on this case in a book on race. The answer is that in calling forth the Ute data to show that selection has operated to exaggerate a "genetic" trait in a single human group Lorenz has crossed the line from explaining a *pan*-human behavioral trait to explaining behavioral differences among human groups on the basis of genetics rather than culture. To accuse the Ute Indians of being genetically more aggressive than other human populations is not only wrong, it is racist. As we shall see below, rather than being an isolated case this is typical of an important aspect of Lorenz's thinking.

Allow me to return now to Ardrey's *The Territorial Imperative,* an extension of the Lorenz-Ardrey aggression-territory hypothesis. Ardrey's argument runs something like this: For at least some animals, territorial behavior appears to be innate. Since territorial behavior appears in many species, it must be adaptive and is, therefore, an essential element in evolution. Territoriality exists in so many species that it must also exist in humans (as a genetic trait). Finally, any behavior in human groups that is in any way connected with property or territory is evidence for the instinct in humans. Evidence against universal territoriality is taken as either incorrect or a momentary aberration of biological structure and is, therefore, detrimental to human survival. Thus, while Lorenz sees evil in the *not incurable* incompatibility between biology and culture, Ardrey sees the territorial drive alone as essential, not only for an explanation of human existence, but also as the major raison d'être of human kind. Schools of behavioral analysis that condemn either the instinct concept or our lust for territory are playing a wicked game because, says Ardrey, they inhibit that which cannot be inhibited and lead us along a biologically dangerous political path.

For Ardrey the territorial instinct is exclusive of, and takes precedence over, the sex drive. In fact, he uses it as the primal explanatory principle for most animal behavior. Even aggression is secondary to territoriality and acts in its service.

Ardrey must account for the fact that territorial behavior takes different forms. He defines it, therefore, as an instinct with an open program! The instinct must be satisfied, but the behavioral path to satisfaction is determined by learning and, in the case of humans, by culture. In practice, however, Ardrey allows fewer degrees of freedom for the expression of this "instinct" than the data demand. He ignores, for example, the fact that the most technologically primitive societies, those that depend on hunting and gathering, are the least territorial of all human groups. In fact, anthropological data shows that private property is the child of culture. It develops into a major preoccupation only with the emergence of complex societies. Allegiance to territory rather than to one's kin is a relatively recent development in human history, along with the invention of the state. If territoriality were "natural" for humans it would also be automatic. Yet, when territoriality becomes imperative for particular human groups in particular historical situations, national conscience is systematically developed through learning, particularly enculturation.

Marriage is linked by Ardrey to the desire for a place of one's own (a territory). He ignores the fact that in many regions of the world, in many different cultures, a place of one's own is far from the mind of either spouse. In fact, the American ideal, neolocal marriage (in which the new couple sets up an independent household), is rare. More common types of marital residence (the newlyweds have to live somewhere) are patrilocal (the couple lives with the father of the groom), matrilocal (the couple lives with the father of the bride), or avunculocal (the couple lives with the groom's mother's brother). Such residence rules are tied to rules of inheritance, social alliances, and sometimes property, but not in any inherent way.

Ardrey bolsters his contention that humans are territorial animals by referring to several unnamed biologists who, like him fail to take culture into account. According to him, one of these experts, when asked if he regarded humans as a territorial animal, replied in the affirmative, noting that signboards all over the country assert that trespassers will be prosecuted. Another replied that humans take property ownership as an inherent right. Taking these statements at face value illustrates a profound failure to realize that what humans consider their inherent rights at any particular historical moment is a product of culture, not biology. Rules against free access to land were met with disbelief by members of many American Indian tribes, and the treaties signed by their chiefs had no meaning for them. The land was there to be used by all, provided that cultural rules associated with it were respected.

Ardrey notes that the basic territorial fact of the human condition was seen by two sociologists, one in the nineteenth century and the other at the beginning of the twentieth. These were Herbert Spencer and his disciple William Graham Sumner, both famous laissez-faire theorists and arch-conservatives. For Ardrey their truth has slipped away under the liberal ethic and the insidious work of the behaviorist psychologists. But even Spencer and Sumner are chided by Ardrey, for neither fit comfortably into the instinct camp. Both (wrongly for Ardrey) felt that the "amity-enmity complex" based on aggression and territoriality was something particularly human and that it could be changed through *social* evolution toward a higher moral order. Both men looked for social causes to explain social events.

Ardrey's own political morality becomes manifest in *The Territorial Imperative* when he predicts that if the Union of South Africa were invaded by white forces, 80 percent of all South African blacks would come to the country's defense. If black forces were to invade, he contends, 100 percent would rise to defend the country. This was merely a hint of what was to come later in his last book, *The Social Contract,* published in 1970. Here we are finally treated to Ardrey's slightly tempered but manifest racism:

> In confronting an environment, the superiority of the individual, of the population, of the *race* at our stage of human history must rest in large portion on the capacity to learn. But the capacity to learn—which we may grant in *Homo sapiens* as more or less equal throughout human populations—is today founded on biological bases varying according to the varying environments in which modern races have evolved. Selective pressures operating on tribesmen living in disease-ridden Africa cannot have been the same as those confronting herdsmen on the wind-swept Central Asian steppes . . . and no variance could have been broader or more anciently rooted than in those geographically isolated groups which we refer to as races. (62)

Interestingly, Ardrey is fully aware of the difference between modern notions about species and populations and himself notes the change in modern evolutionary thinking and classification toward the notion of genetic variation within groups. In *The Social Contract* he underlines this shift away from typological (type specimen) thinking toward the populationist view that species and populations are heterogeneous groups and that genetic variation among individuals in these groups is the normal state of affairs. Thus he specifically cautions against the old notions of uniformity based on the ideal type specimen. Yet he wants to have it both ways. While admitting that populations of a species show genetic variation he wants to argue that Africans are an exception to the general rule:

> The inequality of races, if it exists, must be systematic. It must rest on discernible factors in the differing natural selection placed on the hodgepodge of human mosaics to which we give the term "race." Science possesses today no such discernible factors. We possess only evidence of difference.
>
> In the small black race . . . we have such evidence of superiority of anatomical endowment and neurological coordination that it must be regarded as a distinct subdivision of Homo sapiens. If racial distinction on the playing field is to be accepted, then can there exist theoretical grounds for banishing distinction in the classroom? In the United States the evidence for inferior learning capacity is as inarguable as superior performance on the

> baseball diamond; yet the question of intelligence remains distinctly unsettled. (Ardrey 1970, 62–63)
>
> [The Jensen Report suggesting the genetic inferiority of Negro intelligence is] so persuasive that there were those who could provide no better answer than to threaten Jensen's life. (Ibid., 64)

After agreeing with Jensen and sowing the seed of the IQ argument in the reader's mind, Ardrey retreats somewhat:

> We do not know about race; and that is the final truth today. We know that within a single interbreeding population . . . the accident of the night dictates a diversity of intelligence of such order that between 3 and 3.5 percent must be termed feebleminded. There are between six and seven million mentally deficient Americans. But the occupation of a single geographical area or ecological niche by two populations of differing subspecies, rare in human life, almost unknown in nature, is a man-made situation demanding man-made answers. And until the scientist, without threat to his life, is free to explore in all candor racial differences, and to prove or disprove systematic inequalities of intelligence, an observer of the sciences has little to offer. But then, neither racist nor egalitarian has much to offer, either, beyond emotion. (Ardrey 1970, 65)

Small comfort.

Even before he won the Nobel Prize Lorenz had developed a reputation as a kindly, even grandfatherly, scientist. A famous and widely circulated photograph shows him in water up to his shoulders with a Greylag goose at either ear. In 1972, however, L. Eisenberg (Alland, 1975) found and published an extract of an article that Lorenz had written in German in 1940. This shocking publication proved that, at least during the war, Lorenz supported the Nazi view of racial biology.

> The only resistance that humans of healthy origin can oppose . . . to the penetration of symptoms of degeneracy is based on certain innate schemas. The specific sensibility of our species to ugliness or beauty is intimately tied to symptoms of bastardization caused by domestication which is a menace to our race Generally a man of great value feels strong disgust at the symptoms of degeneracy in members of another race. . . . In certain cases, how-

> ever, we have found a lack of this selective reaction . . . and even contrary to it, an attraction to the symptoms of bastardization. Decadent art provides numerous examples of this change of signs. . . . The enormously elevated reproductive rate among imbeciles has been known for a long time. . . . This phenomenon leads to a situation in which inferior human material is rendered capable of penetrating, and finally annihilating the healthy part of the nation. It is necessary that some social institution foster selection for the qualities of tenacity, heroism, and social utility if we don't want humanity destroyed through a lack of social selection brought about by domestication. (L. Eisenberg; cited in Alland, 1975; translation mine)

Lorenz responded to this revelation with an apology saying that the quote was an error of his youth, that he was no longer a racist, and that politically he was closer to social democrats than to Nazis. In reading a French translation of his book, *Les Huit Péchés Capitaux de Notre Civilization,* however, I came upon the following:

> Defective mutations are to be expected, and sooner or later they are almost bound to occur. If they affect this selfless behavior pattern, they must mean a selection advantage for the individual concerned. We would therefore expect that sooner or later such "asocial elements" parasitic on the social behavior patterns of the normal society members would infiltrate the society. Seen as a whole, the domestic animal is indeed a sad caricature of its owner. In "Part and Parcel in Animal and Human Societies" (1950), I showed that our aesthetic sense of values is clearly associated with those physical changes that occur in the evolution of the domestic animal: muscular atrophy and fatness, with resultant pot belly, shortening of the base of the skull and the extremities, are typical characteristics of domestication that seem ugly in animal and man, while the opposite characteristics appear "noble." Correspondingly, we value intuitively those behavior characteristics that are destroyed, or at least endangered, by domestication.
>
> In "Part and Parcel in Animal and Human Societies," I described in detail the close relation that exists between the threat to certain characteristics of domestication and the values set upon them by our ethical and aesthetic sense. The correlation is too evident to be purely coincidental, and the only explanation lies in the inference that our value judgments rest on built-in mechanisms that intercept certain phenomena of decay threatening mankind. Similarly, it may be assumed that our sense of justice rests on a

> phylogenetically programmed apparatus that prevents the infiltration of society by asocial conspecifics, or members of the species.
>
> A syndrome of hereditary changes, which has undoubtedly arisen in an analogous way and for the same reasons in man and his domestic animals, consists in the remarkable combination of sexual precociousness and persistent youthfulness. Many years ago, L. Bolk showed that man, in many physical characteristics, is much nearer to the adolescent form of his nearest zoological relations than to the adult animals. Arrested development at the stage of youth is known as neoteny. (1973, 94–95; my translation)

As the French say: "Plus ça change, plus c'est la même chose" (The more things change, the more they stay the same).

CHAPTER SEVEN

THREE AMATEURS, PROFESSORS ALL

William Shockley, Michael E. Levin, and Leonard Jeffries

The majority of academics who believe in the relationship between IQ and race are not cranks. Arthur Jensen, for example, is a serious scholar well trained in the area of psychological testing. He may be wrong about race and IQ, but he speaks with professional competence about a subject within the purview of his discipline. There exists, however, another group of academics who are clearly out of their depth when they address this issue. What follows concerns three scholars in this category. They are William Shockley, physicist and Noble Prize winner; Leonard Jeffries, political scientist and former head of the black studies department at City College of the City University of New York; and Michael Levin, philosopher, also at City College.

Shockley, who died in 1989, was a true believer in the IQ superiority of whites over blacks (and of Asians over whites). The sense of (mild) inferiority Shockley felt for his own racial group in relation to Asians is a curious and widely found complex among those who profess (a strong) black inferiority. Professors Levin and Rushton (the latter to be discussed in chapter nine) share this view. The belief that Asians are superior to whites in IQ has been used by some whites as a defense against charges of racism. If one's own racial group is put in an intermediate position, the argument goes, how can one be called a racist? Many Asians in the United States resent this opinion since it stereotypes them, even if, as in this case, in a positive light. Asian students are often expected to do better than others in their classes, a problem that can lead to a poor self-image among those in this group who are not among the highest academic performers. It might also be worth noting that Shockley, Levin, and Rushton ignore the fact that American Indians score lower than whites on IQ tests although they are also of Asian origin. If Asians were genetically superior to whites in IQ then American Indians should be expected to score higher than whites, unless, of course, the deficit was due to environmental factors.

Leonard Jeffries, an Afro-centrist, is a member of that *minority* of black professors who have actively pursued the idea that blacks are superior to whites, not only in IQ but culturally and emotionally as well. Jeffries is difficult to pin down. Almost all of his pronouncements concerning race have been made in class and in speeches presented to various academic and nonacademic audiences. What he has published does not deal directly with black-white IQ differences. It is his public notoriety that got him into trouble with City College.

It has been reported that Professor Levin avoids the IQ argument in the classroom. His ideas on this subject have appeared in professional journals. Several of these are written as specific attacks on affirmative action programs for women and minorities. In general his publications read as philosophical discourse mixed with a good deal of what I consider muddled biology.

Permit me now to discuss Shockley's peculiar place in the IQ argument in somewhat greater detail. As a successful physicist he represents a blatant case of using one's public notoriety to create a forum for personal ideas outside one's professional competence. In the matter of race and IQ he was a rank amateur, yet he was invited to speak about race at some of America's most prestigious universities. Although Shockley had no professional connection to Jensen or any other psychologist specializing in race and intelligence, he greeted the publication of the widely disseminated Jensen Report with enthusiasm and used it, as well as those cited in it, particularly Cyril Burt, to bolster his own views. What was it Shockley believed? First, he believed that IQ and race were linked and that deficits between whites and blacks on IQ tests reflected an inherited deficit in blacks. But Shockley went much further than Jensen. Unlike Jensen he claimed that IQ was absolutely color coded. The average IQ of Negro populations, he said, increased by about 10 IQ points for each 1 percent of added Caucasian genes! Shockley, true to his own logic, was even ready to suggest that blacks could equal or exceed whites in inherited intelligence when the admixture with "white genes" reached 30 or 40 percent! He either knew nothing of independent assortment (one of Mendel's long-standing genetic laws) or believed that the "IQ gene" was closely linked to skin color genes. Of course, skin color is a complex polygenetic trait, and IQ, if it is in any sense genetic, must also involve many genes located at different loci. Shockley's argument for color-coding intelligence is ludicrous and should never have been taken seriously by anyone with even the most rudimentary knowledge of genetics.

Unfortunately, the national press followed Shockley's every word with articles that were almost reverential in tone. For example, *Newsweek* magazine for December 17, 1973, carried an article on Shockley (and Jensen) in their *Ideas* section. It bore the title "The Great IQ Controversy." Citing the polemic generated over IQ, the article takes Shockley's views very seriously while at the same time treating his critics harshly.

> . . . a surprising number of environmentalists have answered with simple contumely. As a well-balanced review of the whole ugly situation in the December issue of *Psychology Today* [incidentally not a professional journal] makes clear, Jensen was denounced by many blacks and leftists as a racist; the editors of the *Harvard Educational Review* and other journals flip-flopped pathetically in the political cross currents, and various professional subgroups of the American Psychological Association and The American Anthropological Association distorted Jensen's findings, denounced his straightforward questions as "obscene" and called for his ostracism.
>
> Why this extreme personal fury? Some of it certainly was a sincere if dictatorial reaction against supplying fuel to the old fire of racial genetics. But the underlying irony may be that these orthodox liberal zealots share the same elitist views as Shockley—which is to say that they fundamentally

> agree with him that a 15 point IQ deficiency, if it were innate would indicate a genuine functional inferiority. (109–10)

Jensen and Shockley *were,* of course, attacked by zealots with little knowledge of the facts in question, but they were also severely criticized by major scholars in the fields of anthropology and biology. Let me reaffirm here that the putative "functional inferiority" of blacks in respect to IQ (cited at the end of the above quote) *was* a major conclusion of the Jensen Report, a conclusion shared in all its details by Shockley. Scholars responding to this conclusion were attacking the entire notion that IQ and race are correlated.

Shockley's views also received coverage in the May 10 and May 17, 1971, issues of *Newsweek.* I believe that both articles bend over backward to present them in as positive a light as possible, although the May 10 article does include a caveat:

> Thus not only did he [Shockley] seem to be somewhat unsure of just what IQ tests do measure, but there was also a suggestion that he is taking IQ tests as an ironclad indicator of absolute intelligence—which even the tests' most vigorous supporters insist they are not. Then there was the fact that in studying the IQ differentials of black sub populations, Shockley seemed to display a knack for picking precisely that group which, because of marginal diet, cultural deprivation and linguistic shortcomings, could be expected to score low on IQ tests even if intelligence were entirely environmental in determination, with no hereditary input at all. (70)

Let us now turn to Leonard Jeffries. I have never had the opportunity of hearing him express his views on race so all I know of him comes second hand from press reports. If these are accurate in content, as I believe they are, Jeffries has turned the race-IQ argument on its head, claiming that blacks are superior to whites. As far as I am concerned, one might call Jeffries's hypothesis a case of "turn about is fair play," even if one disagrees with him. The basic problem is the linking of any "race" to intelligence. It should be noted, therefore, that unlike Shockley or Jensen, Jeffries has been treated with special harshness in the national press. Published attacks on him have bordered on the sensationalism more likely to be found in tabloids and sleazy national weeklies. This contrast becomes even more apparent when we examine press reactions to the published ideas of his City College colleague, Michael Levin, some of whose ideas come as close to science fiction as those of Jeffries.

Time magazine for August 26, 1991, carried an article on Jeffries in its *Nation* section. Under the rubric *Controversies* it bears the title "The Provocative Professor." The subtitle reads, "A black historian draws fire for racist and anti-Semitic remarks, but followers defend his Afro-centric theories." The article goes on to say: "Jeffries serves up outrages of pseudo scholarship that sound sometimes as if they originated in the lodge hall of *Amos 'n' Andy*'s Mystic Knights of the Sea—a rich irony in which Jeffries, a black foe of racism, makes himself sound like the Kingfish, a racist invention of whites" (19–20).

The article also compares Jeffries with his colleague Levin and excuses the lack of furor over the latter's ideas: "If Jeffries were a tenured white professor peddling race hate tricked up as learning would he be more fully criticized? Or less? Another C.U.N.Y. professor, Michael Levin has, outside the classroom, preached the racial inferiority of blacks without being dismissed from his post. Perhaps this is because as a lone academic he speaks for no coherent movement and has no following. Jeffries marches onstage in all the panoply of Afro-centrism" (20).

New York magazine for September 2, 1991, carried an article on Jeffries with the title "The Rise of Afrocentric Conspiracy Theorist Leonard Jeffries and His Odd Ideas About Blacks and Whites." The *National Review* for Sept 9, 1991, covered Jeffries in an article with the title, "A Deafening Silence," and Brian Hecht, writing in the *New Republic* for March 2, 1992, entitled his piece "Dr. Uncool J."

What are Jeffries's main points? As I have already noted, these are difficult to pin down because so far he has published nothing directly on race. According to *New York Magazine* for September 2, 1991, Jeffries said that Jews were responsible for a large proportion of the slave trade between Africa and the New World. He also claimed that Jewish grandees of Spain laid the foundation for the African slave trade even though it is a fact that the Jews were expelled from Spain in 1492 before this trade began. Jeffries is quoted as modifying this point as follows: "We're not talking about most Jews. Most Jews were being beaten up and down Europe, persecuted for being Jewish. We're talking about rich Jews" (33).

The article goes on to detail Jeffries's ideas on black-white differences, primarily the effect of melanin, the chemical in the skin responsible for color, on a series of inherited behavioral traits. Dark skin is claimed to provide blacks with intellectual, physical, and creative advantages over whites. The article quotes Jeffries as saying:

> We've had four melanin conferences. I organized the second one. It's black scientists dealing with this whole question of melanin. Africans developed melanin to protect them from the sun's ultraviolet rays. It's a law. It's not something I picked off my ass. . . . In the caves of Europe you didn't need melanin. . . . Melanin is responsible, based on research I have—one document is 500 pages 2,237 references, from the University of Basil—melanin is possibly responsible for brain development, the neurosystem, and the spinal column. Without these elements we are not human. Whites are deficient of it because of the circumstances of the ice. . . . It appears that the creative instinct is affected. After you [die] . . . I want them to cut and open the brain and see the pineal stems. Twelve stems of the pineal on whites, scientists say, are calcified. (34)

This type of pseudo-scientific theorizing is typical of demagogic arguments that offer "proof" of racial inferiority. Just as the late Senator McCarthy baselessly accused people of communism, no individual scientists and no articles from scientific journals are quoted to back the claim for biological objectivity.

In the *Nation* for September 9, 1991, Jeffries is quoted as having said that Jews and the Mafia had engineered "a financial system of destruction of black people. . . . It was calculated" (252). Jeffries is also quoted as having referred to Diane Ravitch, a U.S. Education Department official, as "a Texas Jew" and a "debonair racist."

Although I have found no research publications by Jeffries concerning these issues I have read a long book review by him that appeared in the *Journal of African Civilization* for June 1986. This article links Jeffries's ideas on race to his Afro-centric position that it was the African continent that gave rise to many if not all of the advances attributed by Europeans to the Greeks. In this review Jeffries presents a highly positive view of *Civilization or Barbarism* by C. A. Diop, an Afro-centric theorist. There are some astonishing things in this review. For example, Diop presents a monogenic theory of human origin in Africa. This view, taken alone, is not necessarily wrong. It is, in fact, the subject of a lively and current debate in physical anthropology and has recently gained favor with most physical anthropologists as the fossil record from Africa, and more recently Europe, has produced more and more evidence (see chapter four). The argument proper, however, concerns the hypothetical *monogenesis* of the human species in Africa versus polygenesis for *Homo sapiens* with the evolution of separate races from *erectus* forms, distributed in Africa, Asia, and Europe. But the argument for the emergence of *Homo sapiens* in Africa says nothing about the color of these ancestral forms, and it does not concern the *color* (much less the race—the two are not identical) of the "original" human being. That is beyond question unknowable, nor does it have much, if anything, to do with the rise and spread of human culture or the relative intelligence of human populations.

Like most Afro-centrists, Jeffries accepts Diop's argument that unlike Europe, pan-African social structure is matrilineal and that this social form favors equality between the sexes. In point of fact there are two matrilineal areas in contemporary Africa. One is a large zone cutting across the lower middle of the continent. The other is a rather small area in the Guinea coast, particularly in what is now the Ivory Coast and Ghana. The rest of the continent is primarily patrilineal. In any case, matrilineal systems in no way guarantee equality of the sexes or superior positions for women.

Jeffries gives high marks to Diop for his scheme of social evolution borrowed (if in distorted form) from outmoded nineteenth century theorists. Jeffries says that Diop:

> . . . makes a major theoretical breakthrough in analyzing history from a non-European perspective and establishing the analytical value of the African-Asian world view. He points out the limitations and errors of Marx and other European theorists who were not able to thoroughly analyze African-Asian societies. Marx characterized African-Asian formations (M.P.A.) as "ephemeral," transitory societies outside of the pale of historical revolutionary movement. (Jeffries 1986, 157)

While it is true that Marx put the "Asiatic mode of production" outside of his evolutionary scheme for Europe, he did so not because he considered it "ephemeral" but, on the contrary, because he felt it was a social form that, once evolved, tended to remain stable through time and, therefore, lay outside his dialectical scheme of economic-political evolution.

Jeffries, apparently unaware of anthropological theory concerning the relationship between the incest taboo and origin of human social organization in its most primitive form, agrees with Diop that the incest taboo was the foundation of the "clan system" and that "clan organization is founded on the Incest taboo which marks the beginning of civilization." Jeffries quotes Diop:

> . . . in the Clan, man is no longer a simple biological animal. He must regulate his sexual and social relations by very strict rules. As a result, Clan formation developed clear notions of parenting, property, inheritance, and individual and group responsibilities. Marrying outside of the Clan, exogamy produced neighboring clan relations and led to a sense of ethnicity and tribe. Environment played a major role in determining whether a Clan became patrilineal (father's lineage) or matrilineal (mother's lineage). The nomadic environment inevitably produced a patrilineal system while a sedentary agricultural environment produced a matrilineal system. Diop states that the division of work at the Clan stage produced a process of social stratification and primitive accumulation which engendered a "clan" system and a tribal structure. As the tribal structure expanded and became complex it developed processes that led to monarchy. (Jeffries 1986, 154)

The propositions offered above are a mixture of oversimplification and misunderstanding of what is known of human social organization. Many societies have been documented as having social organizations simpler than the clan system. These, as do most documented cultures, *have* marriage rules *and* one or another form of the incest taboo. In fact the incest taboo is *universal* in human societies, occurring in different forms everywhere. Contrary to Jeffries's and Diop's ideas, most simple societies (including those below the clan level of organization) have clear notions of parenting, property in some form, inheritance rules, and individual as well as group responsibilities. It is also common for them to have a "sense of ethnicity" (i.e., who they are in opposition to other human groups).

On the other hand, the so-called *tribal* stage of social evolution included by Diop in his scheme of universal social evolution may well be an artifact of colonial expansion. As Europeans spread over Africa and the New World, they often legitimized their conquests by signing treaties with "tribal chiefs," many of whom were created whole cloth by the very same colonial authorities (cf. Fried 1975). While the environment may indeed play an important role in the development of both patriliny and matriliny, and while it is true that nomadic herding populations are exclusively patrilineal, it is not the case as Diop claims that matrilineal social organization can be attributed simply to settled agriculture. Too many counter-cases exist in the ethnographic literature, *including* societies in Africa.

Diop's claim that "the division of work at the Clan stage produced a process of social stratification and primitive accumulation" may or may not be correct. This notion, however, is too schematic to be of much use in understanding how these changes occurred. The same is true for the vague statement "As the tribal structure expanded and became complex it developed processes that led to monarchy." As Diop should know, since he is an Africanist, "monarchy" is a complex concept encompassing a range of political organizations. In sum, by praising Diop's Afro-centric theories, Jeffries reveals his own *very modest* knowledge of human physical *and* social evolution.

On September 19, 1991, the City College Senate voted to condemn Jeffries's remarks concerning Jews. Their complaint was based on things Jeffries had said in his classes at the university that the majority of Senators felt were racist in content. On October 27 of the same year the City College's Board of Trustees voted to give Jeffries a one-year extension as chairman of the black studies department. On March 23, 1992, the university's Board of Trustees voted to remove Jeffries as head of the black studies department. Jeffries challenged this decision in Federal District Court with the claim that the university had violated his freedom of speech. The court sided with Jeffries, awarding him $360,000 in damages, and he was reinstated as head of the department.

The university appealed the decision with the argument that Jeffries had disrupted the operations of City College. The original decision of the court was upheld in April 1994. Using a then-recent ruling of the Supreme Court of the United States, the Attorney General of New York appealed the case to the Supreme Court, which, in turn, ruled that the

Court of Appeals should reconsider its finding. Finally, in April 1995, the appeals court reversed its original decision, but Jeffries remained at his post as a tenured professor.

I shall now turn to Michael Levin, first comparing the reaction within the academy to Leonard Jeffries on the one hand and Levin on the other. It has been personally shocking to me that some have suggested that Levin's ideas on race stand on firmer ground than those of Jeffries. A close look at Levin's arguments, however, particularly when he interprets biological data and theory, reveals weaknesses, including at least one very silly argument that I can only attribute to sloppy, preconceived, thinking.

Philip Gourevitch, bureau chief of *Forward,* writing principally on Jeffries in *Commentary* for March 1992, said the following concerning Jeffries's and Levin's rights to express their views:

> The problem of Michael Levin has shadowed Leonard Jeffries for years, ever since Levin began publishing his belief that blacks are genetically less disposed to intelligence than whites, and that white store owners, for example, are rationally justified out of fear of crime in discriminating against blacks simply on the grounds that they are black. Such theorizing is something of a hobby for Levin; the students in his philosophy classes when polled by the New York *Times* last year expressed unawareness of his fascination with race. . . .
>
> Although [Levin] speaks as a professor of philosophy, he does not claim to speak for philosophers in general, as Jeffries claims to speak for blacks, and he has no apparent following. . . .
>
> Levin is a vigorous supporter of Jeffries's right to say whatever he likes, but Levin and Jeffries are not, in fact, the same thing. . . . Jeffries' classroom teaching has drawn complaints, but he has been exonerated, while Levin, whose students were unaware of his controversial views, had to sue Harleston [president of City College] to get an injunction against the college to prevent further stigmatization and damage to his academic reputation. (Gourevitch 1992, 34–35)

I have no personal knowledge of the relations between Levin and Jeffries, and I applaud Levin's support of Jeffries's right to self-expression, although I would question Jeffries's as well as Levin's professional competence to discuss the problem of race and IQ. Since I have already discussed Leonard Jeffries's views on race I shall now discuss those of Michael Levin.

Levin's favorite forum for airing his views on race (as well as gender) is the *Journal of Social, Political, and Economic Studies* (*JSPES*). More popular articles have also appeared in *Commentary*, a conservative Jewish monthly. The articles I have found bear the following titles: "Race Differences: An Overview" (*JSPES*); "Equality of Opportunity" (*Philosophical Quarterly*); "Why Homosexuality is Abnormal" (*The Monist*); "Comparable Worth: The Feminist Road to Socialism" (*Commentary*); and "Feminism, Stage Three" (*Commentary*). Levin expanded his views on race again, this time in a book, *Why Race Matters*. The criticism of Levin I shall now offer is based, for the most part, on his 1991 article "Race Differences: An Overview." It is my judgment that this paper represents a fair summary of Levin's ideas on race and intelligence, which have, to my knowledge, not changed since.

The article begins as a philosophical tract apparently meant to establish Levin's expertise in that discipline. After citing Hume on the problem of evil, Levin calls forth the null hypothesis to support his views on race and IQ. The null hypothesis simply stated says: "We shall begin by accepting the fact that *no* causal relationship exists between factors A and B." If we wish to *establish* such a relationship we must attack the null hypothesis and prove it false. Levin applies this reasoning to the supposition that there exists a relationship between what most people consider to be *racial* identity and IQ. He says *correctly* that observation contradicts the null hypothesis. He notes that, for *whatever reason*, the *phenotype* black does less well on IQ tests than the *phenotype* white. Here we are dealing only with a question of fact, not interpretation, although I will shortly criticize Levin's use of the word "phenotype" in this context. What is immediately important is the fact that nothing is said about the null hypothesis that shows IQ and racial (as biological) identity to be causally related.

Remember, *phenotypes* are the result of both environmental influences and hereditary factors. Remember, too, that we have *no* sure measure of heritability for IQ in blacks. Using the word "phenotype" under these circumstances is far from innocent. It implies that at least a part of the proved difference in IQ is based on genes: *that part* of the phenotype that is the genotype. If Levin does not want to say genetics is implied in racial differences he should avoid the term "phenotype." He should instead say that the social group referred to as "black" or "African American" consistently scores lower on IQ tests than do members of the social group referred to as "white." On the other hand, if he *does* wish to test the proposition that IQ and race are correlated in a biologically causal way he needs to test a different null hypothesis. The proper null hypothesis would state that there is no biologically causal relationship between IQ and membership in a *biological* population. *This* null hypothesis has never been disproved.

To be absolutely clear let me state this another way. Any null hypothesis must deal with *real* variables. Since I have already shown that biological race is a false category (or false variable), Levin's null hypothesis can be neither proved nor disproved. It is based on a false premise. The fact that blacks in American culture can be defined for purely sociological reasons as a *sociological* race, a group defined by its social, economic, and cultural status in the context of a larger population, does not allow for Levin's null hypothesis as he states it. We need to be rigorous in choosing what variables we include in the testing of any null hypothesis or we risk establishing a real relation between variables that are different from the ones we believe we have chosen.

Levin next attempts to show that IQ tests measure such "independent" variables as "grades, income, job status, job performance, and ordinary judgments of intelligence." "If," he says, "IQ were a test artifact, grades would be no more predictable from IQ than from the sum of the digits of students' telephone numbers" (197). I have no quarrel with the statement that IQ tests indeed measure the qualities mentioned by Levin. Such tests *were,* in fact, *developed* to measure just such types of performance. They were *not,* however, developed to measure the *genetic* component of intelligence. And to admit that the factors noted by Levin are predictable from IQ scores *in no way* implies that any of these factors are *independent* variables. In fact low scores on IQ tests, low grades, poor job status, etc. may all be mutually *dependent* variables all caused by the effects of a *different independent* variable, *social discrimination.* Levin attacks his critics on this point in the following way:

> One would think it obvious that the IQ data concern differences in phenotypic intelligence, but in my experience egalitarians *immediately* interpret

> the data as concerning genotypic intelligence and *immediately* attack the data under that confused and uncharitable interpretation. The polemical advantages of so doing are obvious. Genotypic variance is a (rather natural) *explanation* of some of the phenotypic variance, and the evidence for any explanatory hypothesis *must* be weaker than the evidence for the datum to be explained. Conflating genotypic and phenotypic issues thereby makes the datum of race difference itself appear speculative and uncertain. (Levin 1991, 198)

Yes Professor Levin, phenotypic intelligence *is* what you get no matter what other factors combine to produce the phenotype! But this does not answer the criticism I have offered above.

Levin refers to the "tendency to develop learning ability" as genotypic intelligence. "Indeed," he says "most people use 'intelligence' to mean 'what is innate, not environmentally mediated.'" Most people, of course are not trained to make scientific judgments of this type and, therefore, accepting such a conclusion as a basis for logical scientific argument is, to say the least, highly dubious.

If the null hypothesis concerning the *genetic* cause of IQ deficits in blacks has not been disproved we must remain faithful to it. Additionally, given what we already know about environmental effects on IQ test results for whites and blacks, we should at the very least tentatively accept the falsification of the *other* null hypothesis, that is, that environment has nothing to do with mean IQ scores for groups. So far the environment is a better candidate for influencing group means in IQ. Additionally, it does not seem outlandish at all to me to blame human agency for an IQ deficit in blacks in the light of the history African Americans have suffered, and continue to suffer, in the United States.

Levin claims that once genotypic intelligence has been identified we can then empirically demonstrate its racial distribution. If this appears logical let us remember that there is no such thing as *genotypic* intelligence! *No* organism is a sum of its genes and no genes are expressed directly as a genotype. To be charitable to Levin let us suppose that what he means here is the *heritability* of intelligence. Remember, heritability is a measure of genetic contribution to the variance of a trait IN A GIVEN ENVIRONMENT. Because we lack the data on the heritability of intelligence in African American populations we can make no judgment concerning what Levin incorrectly refers to as "genotypic intelligence."

Levin apparently believes that the genetic problem is solved because data does exist on identical twins reared apart and together, showing IQs of greater similarity than IQs of fraternal twins reared apart and together. After this claim Levin states: "This conclusion is reinforced by the marked depression of IQ with inbreeding, which suggests the control of low intelligence by a recessive allele" (ibid., 200). This is pure nonsense. The factors that might depress intelligence, particularly through inbreeding, have no relationship to whatever genes *might* produce high intelligence. Inbreeding *may* lead to a vast array of deficits caused by recessive alleles combined in greater frequency among related individuals than in a population strictly obeying an incest taboo. However, these deficits are related to pathological conditions that only *indirectly* affect intelligence. It is not appropriate to use defective alleles to prove a point about nondefective alleles. Finally it is highly unlikely that whatever genes may be implicated in intelligence are single alleles. Intelligence is a very complex trait, and if genetics is involved it must be so involved at many loci.

Although he spends a good deal of time at the beginning of his article arguing about phenotypes and intelligence, Levin shifts later to a harder position, saying that "a culture is a product of the aggregated genotypes of its members mediated by their physical environment. There are no other independent causal variables" (ibid., 202). This leads him to warn that a culture's technological level is highly correlated with the level of that culture's individual members. Africa, he tells us, has not produced the number of geniuses equivalent to Asia and Europe. No matter that it is impossible to equate the number of geniuses in a population with its cultural level. Anthropologists have shown that technological innovation in cultures is the result of complex processes involving environmental stimulation, individual inventiveness, historical moment, and borrowing from other cultures! We have no idea how many geniuses any culture contains or has contained at any point in time. One can put this problem into perspective by asking whether or not Darwin would have been an outstanding biologist in this epoch of molecular biology. Darwin was a great *naturalist.* It was his position in this now almost-outmoded field of biology that led him to his theory. Would he have done as well in DNA research?

On page 202 Levin switches to speculating about what physical traits might cause low IQ in blacks. Here he looks as silly as his colleague Leonard Jeffries. He suggests that low IQ might be due to the "more conical Negroid skull," which he attributes to a climatic adaptation. On the next page he cites J. P. Rushton (to be discussed in a later chapter) as having data to show that "Negroids, in comparison to Caucasians, are more violent, more prone to crime, more excitable, more impulsive, more extroverted, less sexually restrained, less inclined to follow rules, less cooperative and less altruistic" (ibid., 203).

As an anthropologist who has done cultural field work in Africa let me say that only one of the characteristics cited by Rushton—more extroverted—applies to the group I studied (the Abron of the Ivory Coast). From my limited experience this aspect of African culture is best reflected in the curiosity and friendship displayed to visiting foreigners, black and white, providing they do not overstay their welcome. But let me also remind the reader that Africa is a vast continent with thousands of different cultural groups. The types of personality characteristics cited by Levin are highly correlated with culture, not with genes. African societies, and the individuals who make them up, are as likely as any other societies and individuals in other parts of the world to vary in these traits.

Levin's actual point is similar to that offered by Herrnstein and Murray to be discussed later. It contains the warning that if blacks continue to outbreed whites in American society, the mean intelligence of the American population will fall to dangerous levels. Levin rejects all environmental explanations for IQ differences by claiming that: "As always the question is why, and, as always, environmental explanations are circular. Slum conditions cannot explain black crime, since 'slum conditions' is just a name for phenomena that include black crime" (ibid., 208). Sure, but the environmental complex under which the majority of African Americans were forced to live in the past, and are still forced to live under in the contemporary United States, presents a valid alternate hypothesis to explain crime rates, which we do know vary according to overall economic situation, class, and economic opportunity.

Levin attempts to protect himself from the charge of being a *vulgar* racist (he admits that he is a "racialist") by claiming that while blacks are genetically inferior to whites in IQ, whites, in turn, are inferior to Asians.

Yet, he calls forth totally unproved adaptations through genetic change to different environmental conditions to explain behavioral differences among races. "In that sense there may be nothing 'abnormal' about black levels of altruism or aggression, if they express genotypes expressed adaptively during Negroid evolution. What complicates matters is that what may be 'normal' Negroid cognitive and social functioning is perceived rightly as 'abnormal' for Caucasoids, just as eating fish is 'normal' for cats but not for horses. It is difficult to appreciate that criteria of normality may differ by race" (ibid., 209–10).

In this quote Levin jumps from species differences between cats and horses to *putative* racial differences within species. In so doing he pushes the outlandish assumption that environmental pressures all over the vast African continent, operating on thousands of cultures and populations, were the same throughout the evolutionary process. No matter that this strains all the evidence cultural anthropology and physical anthropology have amassed over the last hundred years. Levin presents a parody of how evolution works and simplifies beyond recognition the complex differences that exist among African societies (a cultural category) and African populations (a biological category).

Ironically, at least for this reader, Levin concludes by pleading for value neutrality, which he says "is desirable." Even here, however, he immediately adds "by common consent the consequences of the race differences are disturbingly problematic" (ibid., 213).

In general Levin speaks to a limited audience of the convinced. The journals in which he publishes are highly specialized and have a small readership. Unfortunately, his theories concerning race and IQ have reached a much wider audience because they have been picked up by the popular press, with little understanding of the scientific issues under consideration. The fact that two controversial professors at the same university share the same basic hypothesis about race and IQ, and yet disagree about who—whites or blacks—are members of the superior group and who are members of the inferior group, make the controversy between the two men good material for popular consumption. This is particularly the case for a society in which too many are too ready to accept any biological argument concerning race and IQ providing that they fall into the superior group.

One last word about Levin: I agree with him that our educational system should go as far out of its way as possible to treat each student as an individual. There should be nothing controversial about this proposition. But this does not make the proposition that IQ and biological race are linked any more valid or acceptable.

What the three scholars discussed in this chapter have in common is their tendency to poach on areas for which they have neither training nor apparent expertise. If their reputations had not been built in other areas, and/or they had no bully pulpit from which to express their rather bizarre views, they would probably go unnoticed in both the press and by wary academics. Speaking for myself, given the notability of the three, I feel obliged to answer their arguments. It has been said that a little knowledge is a dangerous thing. So is the misuse of one's reputation in a particular field to make claims to knowledge in another. The three cases discussed here differ in that, as far as I know, neither Shockley nor Levin brought their views on race into the classroom. Jeffries, on the other hand, does teach these ideas to undergraduate students, most, if not all, of whom are black. In my opinion this does a disservice to the program he used to direct and to the students who are exposed to his extreme views. The fact that the press has generally made fun of Jeffries, however, does little to combat his particular brand of racism. There is a danger in at least a small number of black studies programs that whatever views are expressed by professors are taken as gospel by students who themselves live in a self-segregated academic situation. Responding to racists demands clear scientific arguments, not derision, no matter how outlandish the ideas professed appear.

CHAPTER EIGHT

HARVARD IN THE ACT

Richard J. Herrnstein, Charles Murray, and Meritocracy

In September of 1971, *Atlantic* magazine published an article by Harvard psychologist Richard J. Herrnstein. Avoiding the race and IQ issue, the author argued instead that IQ and social standing were closely linked in American society. The lead, presented in bold type just under the title and the author's name, contained the following propositions, followed by two check boxes marked true and false: "1. If differences in mental abilities are inherited, and 2. If success requires those abilities, and 3. If earnings and prestige depend upon success, 4. Then social standing will be based to some extent on inherited differences among people."

What follows is a carefully argued and, if not read closely, moderate-sounding defense of the proposition that IQ and social standing are highly correlated. Herrnstein believes that this linkage has important consequences for the future of American society. The early part of his argument resembles Levin's in one important respect. Both authors note that members of high IQ groups are far more successful professionally than those in moderate to low IQ groups. There is, of course, some truth in this statement. So, it would seem, what is important here to them is *not* what *causes* high IQ. Instead, what is treated is the proposition that IQ is an independent variable leading, other things being equal, to professional success. While this proposition might appear to differ from the usual arguments concerning IQ and race it is actually a diversion that deflects attention from the crucial social issues involved in IQ studies today.

Let us follow the logic of Herrnstein's argument in his *Atlantic* article. He begins by saying that in the United States " . . . there is a powerful trend toward meritocracy—the advancement of people on the basis of ability, either potential or fulfilled, measured objectively" (45). We are then told that this trend has lately been deplored by those who object to the IQ test's use as entry into the establishment. The issue, he says is "intensely emotional."

> Yet should not the pros and cons be drawn from facts and reason rather than labels and insults. For example, is it true that intelligence tests embody only the crass interests of middle America, or do they draw on deeper human qualities? Is the IQ a measure of inborn ability, or is it the outcome of experience and learning? Can we tell if there are ethnic and racial differences in intelligence, and if so, whether they depend upon nature or nurture? Is there one kind of intelligence, or are there many, and if more than one, what are the relations among them? . . . For those who have lately gotten their information about testing from the popular press, it may come as a surprise that these hard questions are neither unanswerable nor, in some cases, unanswered. The measurement of intelligence is psychology's most telling accomplishment to date. (45)

Herrnstein goes on to provide a thumbnail sketch of the history of IQ testing since James McCattell in the United States at the end of the nineteenth century. He lingers fondly over Alfred Binet, not quite willing, however, to admit that Binet never saw IQ as a measure of innate abilities. Instead Herrnstein compliments Binet for having "finessed the weighty problem of defining intelligence itself. He had measured it without having said what it was" (46).

Herrnstein gives cultural relativism its due when he admits that IQ tests are culture bound rather than culture free. However, he manages to convert this limitation into an advantage, for he considers IQ tests to measure how successfully children negotiate their own cultures!

This argument makes perfect sense. However, if IQ test responses are taken by scholars to be culture bound and, therefore, a function of environment, to be consistent the same scholars would have to reject the putative *genetic* basis of *group* differences in performance on these tests. You just can't have it both ways. Yet, as we shall see below, Herrnstein takes the genetic approach to explaining group differences very seriously.

Herrnstein poses the question: Is intelligence one thing or a complex of *different unrelated* traits? For a tentative answer he turns to the British psychologist Charles Spearman. Spearman developed a statistical tool for testing the proposition that the qualities displayed by answers to different categories of questions on IQ tests reflected a single factor that he called "g," for general intelligence. This concept was later refined by L. L. Thurston, who divided Spearman's g into a set of primary mental abilities. These included spatial relations, perceptual ability, verbal comprehension, numerical ability, memory, word fluency, and both inductive and deductive reasoning. Herrnstein argues that Thurman successfully managed to intercorrelate some of these primary mental abilities, (verbal skills vs. mathematical skills, for example) possibly with the elusive g at the top of the system. (1971, 49)

Herrnstein then returns to the function of IQ tests, which, he says correctly, is to predict success in school, occupational suitability, and intellectual achievement. But, he admits, school performance is affected by environment more than IQ. IQ, he says, is necessary for, but not sufficient for, good schoolwork. Speculating about what else is needed Herrnstein suggests that interest, emotional well being, and energy might fill the bill. These, he says, are also required for success in business. Moving toward his theory of meritocracy Herrnstein notes that IQ and social class correlate well. The upper class scores on the average about 30 points above the lower class. But, he warns, this does not provide a basis for concluding that no poor people have high IQs. Instead he looks at an average ranking of IQ for 74 occupations. He finds that IQ affects one's occupation and occupation affects one's social standing. Therefore, he concludes, IQ affects social standing.

Herrnstein next turns to a forty-year study undertaken by Lewis M. Terman and his colleagues at Stanford University. This group followed the lives of a selected sample of gifted people with IQs in the 150 range. Herrnstein notes that there was an excess of children whose fathers were professionals, particularly disproportionate to the number of children whose fathers were laborers. Additionally he remarks that the sample contained an excess of European Jews and a shortage of Latins, Eastern Europeans who were not Jewish, and Negroes.

Again Herrnstein's interpretation of this data assumes that IQ is the independent variable and occupation the dependent variable. But to accept the data at face value also implies that Northern and Western Europeans as well as Jews are genetically superior in intelligence to Latins, non-Jewish Eastern Europeans, and blacks. Thus, an ethnic distinction likely to be caused by social conditions is turned into a hidden racial hypothesis. To unscramble the facts we need to know at least the following: 1. Were there differences in "need achievement" between the over- and under-represented groups? 2. Were there differences in time of immigration to the United States between the over- and under-represented groups? 3. Were their differences in English language skills for both the parents and the subjects tested?

More general criticisms of this work have been offered. In his book on race, *The Mismeasure of Man,* Stephen J. Gould treats Terman's data with skepticism.

> Terman investigated IQ among professions and concluded with satisfaction that an imperfect allocation by intelligence had already occurred naturally. [In this sense he is the father of meritocracy.] The embarrassing exceptions he explained away. He studied 47 express company employees, for example, men engaged in rote, repetitive work "offering exceedingly limited opportunity for the exercise of ingenuity or even personal judgment" (1919, p.275). Yet their median IQ stood at 95, and fully 25 percent measured above 104, thus winning a place among the ranks of the intelligent. Terman was puzzled, but attributed such low achievement primarily to a lack of "certain emotional, moral, or other desirable qualities," though he admitted that "economic pressures" might have forced some "out of school before they were able to prepare for more exacting service" (1919, 275). In another study Terman amassed a sample of 256 "hobos and unemployed," largely from a "hobo hotel" in Palo Alto. He expected to find their average IQ at the bottom of his list; yet, while the hotel mean of 89 did not suggest enormous endowment, they still ranked above motormen, sales girls, firemen, and policemen. Terman suppressed this embarrassment by ordering his table in a curious way. The hobo mean was distressingly

> high, but hobos also varied more than any other group, and included a substantial number of rather low scores. So Terman arranged his list by the scores for the lowest 25 percent in each group, and sunk his hobos into the cellar. (Gould 1981, 182–83)

Reflecting on Terman's conclusions Gould goes on to say: "Had Terman merely advocated a meritocracy based on achievement, one might still decry his elitism, but applaud a scheme that awarded opportunity to hard work and strong motivation. But Terman believed that class boundaries had been set by innate intelligence. His coordinated rank of professions, prestige, and salaries reflected the biological worth of existing social classes" (ibid., 183).

After citing Terman's study Herrnstein moves approvingly to the Jensen Report. He says: "The article is cautious and detailed, far from extreme in position or tone" (1971, 55). He uses Jensen's material on twins reared apart, while recognizing that these data are, in fact, not appropriate for deriving a heritability figure for blacks.

Herrnstein is also aware that whatever the heritability figure for IQ, even within the white population at a given time, changes in environmental variables can bring about sharp changes in the average phenotype of any population.

Intelligence, he admits, may be drifting up or down for environmental reasons from generation to generation, notwithstanding its high heritability. Average height, he reminds us, is said to be increasing in the population—presumably because of diet and medicine—even with its .95 heritability, but he neglects to mention that the heritability figure given (.95) is itself dependent on *both* genetics and environment in a *given* population in a *given* environment. I have to suppose that he is speaking here about the American population. The heritability of height is, or was, undoubtedly much lower among pre–World War II Japanese, who were, on the whole, very short when compared to Americans of the same generation. Since the war, changes in the Japanese diet have led to rapid and marked increases in the average height of each generation. As the average Japanese becomes taller the *heritability* of the trait (remember this means the percentage of variation due to genetics) becomes larger since average differences within the population increasingly become a matter of genetics. In other words, what is happening is that the penetrance of the genes for height increases with environmental change.

Instead of explaining the problem this way Herrnstein says: "But if height has changed, why not intelligence? After all, one could argue, the IQ has a heritability of only .8, measurably lower than that of height, so it should be even more amenable to the influence of the environment. That, to be sure, is correct in principle, but the practical problem is to find the right things in the environment to change—the things that will nourish the intellect as well as diet does height" (58).

Well, yes—we should change the environment to increase intelligence, but if we do so that also means that the heritability of intelligence, whatever it might be, will also change. Does Herrnstein know this? The answer is "yes," for on page 63 he says that a more favorable environment for intelligence will increase its heritability!

Concluding his argument, Herrnstein returns to the syllogism introduced at the beginning of his article. Assuming it has been proved true (and this is far from evident) he draws the following conclusions:

1. A better environment for intelligence will mean that intellectual differences will be increasingly inherited.
2. Social mobility will decrease as it is inhibited by increasing innate differences among the brighter and duller segments of the population.
3. This process will increase the social gap between the upper and lower classes, making social mobility ever more difficult. Those left at the bottom of the social ladder will be stuck there.
4. As technology becomes increasingly more complex, unemployment may run in a family's genes.
5. This emerging meritocracy will depend on more than genes for high intelligence. There will also be increasing selection for such other (possibly) inherited traits as "temperament, personality, appearance, perhaps even physical strength or endurance." (63)

Herrnstein's *Atlantic* article, which stirred up a hornet's nest of protest shortly after publication, was reviewed by Constance Holden in the July 6, 1973, issue of *Science* magazine, the official organ of the American Association for the Advancement of Science. In her attempt to provide an objective and value-neutral view of Herrnstein's arguments she ignored most of their objective shortcomings. For example, she said: "The uproar seems to have come as a surprise to Herrnstein, who says he approached the writing of the article as a nonpartisan observer" (Holden 1973, 36).

And instead of analyzing Herrnstein's arguments she makes reference to mutual name-calling between Herrnstein and one of his critics, Leon Kamin at Princeton.

However, Holden did see one implication of Herrnstein's argument, which would to be developed at great length in *The Bell Curve:* "While Herrnstein said next to nothing about blacks in his article, . . . Herrnstein was saying that if blacks weren't making the grade it meant they weren't born with the brains to do it" (ibid., 36).

To be fair to Herrnstein it must be said that a number of his critics generally did ignore or distort at least *some* of his arguments. But Herrnstein's work is open to more reasoned criticism precisely because it contains such a strong set of biases in the direction of the unproven assumption that IQ is largely a matter of genetics. Thus he tends to ignore the limits of heritability studies as they are applied to cross-racial studies of intelligence.

Herrnstein is also open to criticism for his simplistic assumptions about the present and future of American society. Just as he accepted data, more or less without criticism, to show that IQ was inherited, he appears to have accepted an extreme laissez faire view of society. In this respect his views are similar to those expressed by such nineteenth-century "liberals" as Herbert Spencer in England and William Graham Sumner in the United States. The difference, of course, is that Herrnstein is a *twentieth*-century psychologist. Few of his colleagues in the social sciences today, even the conservatives among them, take either Herbert Spencer or William Graham Sumner very seriously.

One of the most reasoned arguments against Herrnstein's data and conclusions was published in the September–October 1972 issue of *Society.* The authors, Karl W. Deutch and Thomas Edsall, note that the Terman data cited by Herrnstein is tainted by the fact that environmental differences among the members of the Terman sample were highly restricted. The majority of Terman's gifted children came almost exclusively from the white middle class. The twin studies fare no better since all four samples were, as I have already noted, drawn from the white population only. More interesting, because less widely included in criticisms of Herrnstein's argument, are a series of points concerning his sociological premises:

> As Herrnstein sees it, vigor, beauty, strength, kindness, patience, sensitivity, perceptiveness and other characteristics of human behavior and personality would not cut across any supposedly hereditary IQ distribution and would not dilute or largely nullify its effects for most families in the course of a few generations. If any such hereditary effects existed, a considerable degree of formal or informal segregation in social contacts and in the choice of marriage partners would be essential if the social effects of the heredity assumed by Herrnstein were to persist over time.
>
> But this assumption implicit in Herrnstein's arguments conflicts with his assumption of equality of environmental conditions and education opportunity. (Deutch and Edsall 1972, 75)

The same authors go on to note Herrnstein's elitist attitude toward society: "In Herrnstein's world, no one does any work for its own sake or for the sake of other people, nor does anyone choose a job because he likes to do it. If ditch digging carried the highest external rewards, says Herrnstein, the ablest individuals would compete for ditch digger's jobs" (ibid., 75).

Although he did not directly address the issue of race and IQ in the *Atlantic* article, Herrnstein finally blew his cover on this issue in 1990, when he published "Still an American Dilemma" in *Public Interest* (1990). This is a long review of *A Common Destiny: Blacks and American Society*, the report of a commission on the status of black Americans. After saying that the report is a serious work that deserves study, Herrnstein tells us that it suffers from one crucial failing:

> In assessing the gaps separating white and black Americans, it obstinately refuses to consider the evidence concerning racial differences at the individual level. On the infrequent occasions when it offers or hints at explanations of the data that it presents, it suggests that the central determinant of black-white relations in America is discrimination—past or present, deliberate or inadvertent by whites against blacks. I do not assert that racial

> discrimination is insignificant or tolerable; I do, however, contend that the study wrongly ignores the evidence at the individual level that I discuss here. (Herrnstein 1990, 4)

Herrnstein considers the IQ argument pertinent because whites and blacks in the United States overlap in every important social and economic measure. He reminds us that there are poor whites just as there are poor blacks. Thus the book: "ignores the alternative model, the 'distributional' model, which explains the overlapping of the populations and their differing averages by referring to characteristics of the populations themselves" (ibid., 6). What Herrnstein means here is that in his view the study ignores the possibility that on the *individual level* the average black may have a lower IQ than the average white (ibid., 6). He then goes on to warn that if the distributional model is correct, then affirmative action programs lose their moral strength, becoming mere products of political judgment.

Herrnstein notes that the gap between whites and blacks has decreased recently. Admitting that the legacy of discrimination cannot be discounted, he nonetheless states that we do not really know where the IQ differences come from and what would be needed to make them go away.

The main point in Herrnstein's review is an attack on affirmative action programs, but he is careful to state the problem in such a way as to appear fair minded. Thus, on page 11 he says: "Given the hypothetical choice of being black with an IQ of 120 or white with an IQ of 80, one should choose the former to get ahead in America now or in the readily visible future." Sure—but note the spread in the choice he gives us. This is a "red herring." Worse in my view, Herrnstein, playing on the side of the angles, warns that affirmative action might cause a severe backlash (anger against blacks) when the facts about race and IQ become known to the public at large. Thus the problem of racial discrimination is turned on its head.

And, Herrnstein warns, the unanticipated consequences of affirmative action will be overwhelmingly bad for society at large. This will occur when the group suffering the effects of the new discrimination (whites, one must suppose) discover that the programs were useless in the first place. Woe unto those who ignore "the iron rule of selection!"

"The objective facts about performance will produce damaging subjective expectations. It may well be that the gains to society from attempting disproportionately to benefit blacks would outweigh the losses resulting from differential performance—if the performance gap is small enough. But does anyone doubt that the costs may outweigh the benefits, if the performance gap is large enough?" (ibid., 17).

All this in spite of the fact that so far no convincing studies have shown that genetics rules over the environment when it comes to group differences in IQ. Beyond this, *no* study exists that puts IQ differences, whatever might cause them, into a range that suggests convincingly that such differences matter in the real social world.

The fall of 1994, the year in which Herrnstein died, saw the publication of a massive 845-page book, *The Bell Curve,* by Herrnstein and Charles Murray, a political scientist who works at a conservative Washington thinktank, the American Enterprise Institute. Combining Herrnstein's views on race and IQ as well as his previously published arguments concerning meritocracy with Murray's conservative agenda for healing America's social ills, the book argues that (1) those at the top of the heap of the business and intellectual world in the United States have risen to their positions because they are simply the brightest members of the population; (2) that this brightness is reflected in IQ scores; (3) that IQ is in large measure caused by heredity; (4) that *racial* differences in IQ are largely inherited; (5) that the average Asian is, by heredity, smarter than the average white, who is, by heredity, smarter than the average black; (6) that such social problems as a rising rate of illegitimacy are linked, at least in part, to IQ differences, with low-IQ women having more illegitimate children than high-IQ women; and (7) that social policies, including welfare, Head Start programs, and affirmative action, increase rather than diminish inequality in American society. Need it be said that such an argument is social Darwinism all over again! Although I take exception to much of what Herrnstein and Murray argue in *The Bell Curve,* because this book deals specifically with race and IQ I will therefore limit myself in what follows to that narrow issue.

The Bell Curve received immediate attention from the popular press. Shortly after publication it was reviewed in *Newsweek, Time, U.S. News and World Report,* and *The New Republic,* as well as in the *New York Times,* and the *New York Times Sunday Book Review* section. *Newsweek* and the *Times Book Review* both featured it in cover stories. While the *Times* review was rather noncommittal, in the weeks that followed an editorial in the daily edition, and several letters to the *Times* were all more negative. Unfortunately, the book's major flaw was overlooked in all of the early critical reviews. It should come as no surprise to the readers of this book that the key error in *The Bell Curve* is a replay of Jensen's systematic *misuse* of heritability as a concept. No matter how many studies Herrnstein and Murray cite showing IQ differences among the races, this error calls their entire enterprise into question. Furthermore, from a close reading of *The Bell Curve* I would question the idea that such an error is due merely to chance or carelessness. My reasons for this skepticism will become apparent in the arguments to follow.

Both Herrnstein and Murray are social scientists and both are trained in statistics. They have to know that heritability is a measure of variance. Herrnstein was, as I have noted above, perfectly familiar with the technical meaning of the term. The way the authors treat this key concept in arguing for genetically caused racial differences in IQ prepares the unwary reader to *not* notice their general bias. They do this by hiding a more or less correct discussion of heritability in a confusing—because self-contradictory—discussion that comes late in the book. Even a careful and intelligent reader might fall into this trap. For example, an otherwise excellent, comprehensive, and critical review by Alan Ryan, professor of political science at Princeton, published in the *New York Review of Books* for November 17, 1994, begins as follows: "The *Bell Curve* is the product of an obsession, or more exactly of two different obsessions. Richard Herrnstein—who died on September 24 of this year—*was obsessed with the heritability of intelligence, the view that much of the largest factor in our intellectual abilities comes from our genes*" (7, italics mine).

Hopefully, readers of this book need not be reminded what heritability really is. Nonetheless, let me remind you that this measure tells us what contribution genetics makes to the observed variation of a trait *only* in a *specific* population living in a *specific* environment. I realize that even now lay readers might find it perverse that the heritability for a trait that is *100* percent genetic but that does not vary (the case I have already used is a population in which everyone is blue eyed) *is* zero! It is, however, a true statement and this is a crucial point. Because it is an environmentally *specific* measure of genetic contribution, heritability figures for one population can *never* be correctly applied to another population. Environmental effects may be responsible for *all* observed difference between populations even when the trait in question has the same genetic component in these same populations. Herrnstein and Murray *do* get around to telling us this, but in a most peculiar way, certainly not with the clarity we have seen in Herrnstein's 1971 article. Under the heading *Heritability and Group Differences* on page 298, they say:

> A good place to start is by correcting a common confusion about the role of genes in individuals and groups. As we discussed in Chapter 4, *scholars accept that IQ is substantially heritable, somewhere between 40 and 80 percent, meaning that much of the observed variation in IQ is substantially genetic.* And yet this information tells us nothing for sure about the origin of the differences between races in measured intelligence. This point is basic, and so commonly misunderstood, that it deserves emphasis: *That a trait is genetically transmitted in individuals does not mean that group differences in that trait are genetic in origin.* Anyone who doubts this assertion may take two handfuls of genetically identical seed corn and plant one handful in Iowa, the other in the Mojave Desert, and let nature (i.e., the environment) take its course. The seeds will grow in Iowa, not in the Mojave, and the result will have nothing to do with genetic differences. (298, first italics mine, second italics theirs)

In the next section on the same page the authors go on to admit that by analogy the environment for American blacks has been close to desert conditions and for whites close to those of Iowa. This is followed with the statement: "We may apply this general observation to the available data and see where the results lead." Peculiarly, the results lead Herrnstein and Murray to conclude that whites and blacks can be compared by using an estimate of the genetic effect on IQ for both groups derived only from one of these groups—whites. It is reasonable to hold this effect constant, they say, for the two groups at 60 percent (in their view a middle ground estimate). So once again the authors make believe that heritability is a *constant* rather than a *variable!* Such reasoning reduces the rest of their analysis to an excursion into science fiction.

On reading *The Bell Curve* one quickly discovers that Herrnstein and Murray hate to be accused of bias. In attempting to prove their lack of race prejudice the authors claim that when it comes to race and intelligence they are "agnostic" but only in so far as determining the *percentage* contributions of genes and environment to average IQ scores (311). More importantly, "Measures of intelligence have reliable statistical relationships with important social phenomena, but they are a limited tool for deciding what to make of any given individual (21). Additionally they say:

> Despite the forbidding air that envelops the topic, *ethnic* differences in cognitive ability are neither surprising nor in doubt. Large human populations differ in many ways, both cultural and biological. It is not surprising that they might differ at least slightly in their cognitive characteristics. That they do is confirmed by the data on *ethnic* differences in cognitive ability from around the world. (269, italics mine)

On one level this statement is a truism; on another level it serves to hide the author's intent. I doubt if any anthropologist or psychologist would argue against the reasonable position that at least some *ethnic* groups are likely to differ *somewhat* in cognitive characteristics. But ethnic groups are *not biological races,* even as Herrnstein and Murray use the term (see their inclusion of Latinos as an ethnic group on page 275). Nor does this statement, as it stands, imply that whatever differences test results might reveal are in any way genetic. Such differences could be due to purely cultural factors such as foci of interest, environmental effects, biases built into the tests themselves, or any of these taken together. If there is any doubt that this is meant to mislead the reader, let it be said that this quote comes at the head of a chapter from a section of text set off from the rest by italics. In their "Note to the Reader" at the front of *The Bell Curve,* the authors inform us that their book is designed: " . . . to be read at several levels. At the simplest level, it is only about thirty pages long." Each chapter presents a précis of its argument and main findings as well as the conclusions. " . . . You can get a good idea of what we have to say by reading just those introductory essays" (xix). It is unfortunate that, when faced with an 845-page book, too many readers will take Herrnstein and Murray's advice and read only these frequently misleading précis.

But there are more serious problems in *The Bell Curve,* all consistent with my suspicion that the authors intentionally engaged in some tortured stylistic tricks to hide or render incomprehensible data that run counter to their arguments. Herrnstein and Murray declare that Spearman's concept of general intelligence (g) is generally accepted by psychometricians. In fact, the putative existence of g is a widely debated topic. Not all psychologists, anthropologists, and biologists concerned with this issue agree that intelligence can be reduced to a single concept or measure.

Although they claim agnosticism concerning the total hereditary contribution to IQ differences, they consistently load the dice in favor of the widest hereditary spread possible between whites and blacks. When studies favoring *environmentally* stimulated IQ increases are cited, and a few are, these are either presented with little or no comment or explained away. Thus Herrnstein and Murray cite C. T. Ramey's fifteen-year Abcedarian Project showing consistent and long-lasting IQ gains (of up to 20 points!) in early stimulation for children coming from stimulus-deprived environments (Ramey 1992). These results are dismissed by claiming that subject and control populations are probably incomparable.

Why this skepticism? Data drawn from experiments in laboratory rats shows that very early stimulation has a major effect on brain development on both the micro and macro levels. These same studies show that maze-running ability is enhanced among stimulated animals when compared to genetically identical controls (see Diamond 1988). It is, of course, dangerous to extrapolate from rats to brain development in humans, but such work does provide a theoretical explanation for Ramey's data concerning, as it does, early environmental stimulation and enhanced behavioral performance.

I have read a description of the Abcedarian project and find it to be a theoretically sound, well-designed, and carefully researched longitudinal experiment. Its goal is to determine the effects of early and continuous stimulation via a preschool enrichment program on children beginning at three months of age and continuing through the preschool years. Although begun in 1972, the subjects of this project and a matched control group are still being followed by the research team. The study reveals significant and long-lasting IQ gains for subjects.

Additionally it shows that parental participation in the project has had an important effect on siblings of the experimental group as well as on their mothers, who were more likely to find employment and to continue their education than mothers of children in the control group (Ramey, MacPhee, and Yeates, 1982).

In contrast to their skepticism concerning the viability of environmentally stimulated IQ increases, when Herrnstein and Murray present data showing longitudinal increases *and then decreases* in the IQ of subjects exposed to environmental stimulation, these studies are invariably put in a favorable light. For example, in discussing Project Head Start, Herrnstein and Murray say without further comment that IQ gains found in youngsters just after leaving the program rapidly disappear as they move into "mainstream" schools (304) but offer no hint that conditions in what are often in effect *ghetto* schools might well depress scores.

Of a study of children of white, black, and mixed racial ancestry adopted into white households in the state of Minnesota, Herrnstein and Murray note initial IQ differences showing that adopted black and mixed children were above the average for blacks in the United States. The mean IQs were 117 for the biological children of white parents, 112 for white adopted children, 109 for adopted children with one white or Asian parent and 109 for children of black parentage.

Here Herrnstein and Murray neglect the possibility that *black* children (in the United States this means the children of two or one parent defined as black) brought up as adoptees in white households do not experience their environment in the same way as white children in the same households might. I would suggest that black children adopted by whites are probably subject to special stresses associated with their position both within and outside the household. This stress might well explain depressed IQ scores.

Suspected flaws in the cited research are documented in *Measured Lies: The Bell Curve Examined,* by J. L. Kincheloe, S. R. Steinberg, and A. D. Gresson III, published in 1996. These authors point out that that the researchers failed to account for two factors: pre-placement history of the adopted children and the effects of attrition on overall results (when children have dropped out of the study sample and are thus lost to further measurements). No data of either type were presented by the authors of the original study. The lack of such data calls into question the validity of the results.

Additionally, the *only other* study cited in *The Bell Curve* that sidesteps the heritability problem is one that *directly contradicts* the Minnesota study. This is located in a sidebar to be found on the same page as that which refers to the Minnesota study (304). It concerns the IQ scores of children, the issue of black G.I.s from World War II, and German women. These data reveal *no* significant differences in average IQ between *mixed* black-white children and *white fully* German children. Although the German data is as important to the issue at hand as the Minnesota study, Herrnstein and Murray have already guided their readers around these sidebars by suggesting that these inserts can be skipped as "tidbits" even though they admit that they do add something to the discussion.

It would seem that the German data is either a mere "tidbit" or, as an "alternative way of thinking," it interferes with the message of *The Bell Curve.* This is true only in the sense that it contradicts the arguments of the book.

In a review of "*ethnic*" differences in "*cognitive ability*" Herrnstein and Murray tell us that Askenazi Jews of European origin test higher than any other ethnic group (275). What the authors fail to note is that genetic studies comparing Askenazi Jews with the Poles and Russians among whom they lived show little *genetic* differences among the three groups.

Like Jean Philippe Rushton, to be discussed in the next chapter, Herrnstein and Murray suggest that the average Asian scores from one to three points higher than the average white, but they fail to mention that American Indians, whose genetic roots are in Asia, score consistently lower than whites. These additional data, of course, support the argument that environment rather than heredity is responsible for the observed differences. In the same vein, admitting that Latinos cannot be considered a race (they are really a series of different ethnic groups) and that many Latinos might have language difficulty with IQ tests, Herrnstein and Murray nonetheless say that Latinos score from about a half to one full standard deviation below whites (275). Why do they bother to present this admittedly contaminated data if not to confuse the lay reader?

Herrnstein and Murray estimate the "black population" of the United States to be about 30 million, which is probably correct (278). What they fail to consider is that this "population" is actually *many* populations with different admixtures of African, European, and Asian (Native American) genes. Should we commend them for not suggesting, as did William Shockley, that light blacks should score higher on IQ tests than dark blacks? Additionally, consistency demands that they should have questioned a study showing that the IQ of "colored" students in South Africa is *similar* to that of American blacks (289). Surely they were aware that all nonwhite South Africans suffered until the recent past from a more corrosive system of racial discrimination than even American blacks. Their willingness to load the dice in their favor is consistent with the fact that while they admit the following they hide it in a footnote (footnote 130 of chapter 13): "An earlier study showed no significant association between the amount of white ancestry in a sample of American blacks and their intelligence test scores. . . . If the whites who contributed this ancestry were a random sample of all whites, then this would be strong evidence of no genetic influence on black-white differences. There is no evidence one way or another about the nature of the white ancestors" (729).

I think it bizarre to assume that mostly low IQ whites contributed to the mixed gene pool of African Americans. Of course, such an assumption is a major insult to generations of descendants, black and white, of *white* Southern plantation owners!

An informed reader, familiar with Herrnstein and Murray's sources, will be struck by distortions of the original authors' conclusions, particularly in discussions of race and IQ in other cultures. What follows is just one of many examples.

Herrnstein and Murray cite Kendall, Verster, and Von Mollendorf's study of test performance among African blacks and take this data to show a strong (putatively genetic) deficit placing Africans as low as or lower than American blacks. Here is what the authors of that study actually say in their conclusion:

> In Africa, the psychological literature suggests a wide range of cultural and environmental factors which, in extreme combinations, would inhibit adaptability to a modern, technologically based life style. . . . These conditions impoverish (by western standards) the typical domestic environment of the rural African peasant, and also foster a diet which more often than

> not is rich in carbohydrates but protein-deficient. Education systems throughout Africa are often inadequate and the quality of teaching can vary markedly from school to school. . . . There would thus appear to be many cultural and environmental variables at play in moderating test performance levels; so many, in fact, that one wonders whether there is any point in even considering genetic factors as an additional source of variance between the average performance levels of westerners and Africans. (Kendall, Verster, and Von Mollendorf 1988, 326)

Although more could be said about the data used by Herrnstein and Murray to support their arguments, I am sure that by now the reader is suffering from data fatigue. By way of closing this chapter let me cite a statement from *The Bell Curve* that reveals more than Herrnstein and Murray may want to admit about why the IQ argument comes up at least once every generation. "... [W]e have become convinced that the topic of genes, intelligence, and race in the late twentieth century is like the topic of sex in Victorian England. Publicly there seems to be nothing to talk about. Privately, people are fascinated by it" (297). It does not surprise me that people are fascinated by biological arguments for inequality.

The real and consistent message of *The Bell Curve* is that the rich and powerful in American society have risen to the top on the basis of merit, and that merit itself can be measured in points of IQ. Herrnstein and Murray do hedge somewhat on this issue. They present a ready-made response to those critics who take them to task for arguing that IQ is genetic. They say that even if high IQ is totally *environmental* in origin, the effects on the social system will be the same as if it were genetic: a rising division among the smarter and therefore better-placed individuals and the for-whatever-reason less intelligent who constitute, in growing numbers, the have-nots. This will create a situation of widening financial and social differentiation in the United States. Arguing this way the authors could just as well have avoided the race issue entirely, thus saving them from the charges of racism they so vigorously deny. This begs the question of why genetics is, in fact, dragged into *The Bell Curve.* I think the answer is clear. The emphasis on race and IQ comes from Herrnstein and Murray's noninterventionist political position. They argue consistently against affirmative action and other social programs. If the causes of black-white IQ differences are biological rather than environmental, then their arguments against ameliatory social programs are completely logical. If the environmentalists are correct, then interventionist social programs might help solve what Gunnar Myrdal (1944) so long ago called "the American *dilemma*."

CHAPTER NINE

FROM BEYOND OUR BORDERS

J. P. Rushton and H. J. Eysenck

This chapter examines the work of J. P. Rushton, a Canadian psychologist and professor of psychology at the University of Western Ontario, and the psychologist H. J. Eysenck who lived and worked in England until his death in the late 1990s. Both argue for correlations between race and IQ. While Eysenck's major publication on race (*The I.Q. Argument*, 1971) takes the now familiar route of correlating IQ studies with race, Rushton relies on head measurements of living subjects and, in one case, on data derived from brain scans. He claims to have shown significant relationships between brain size, racial membership, and IQ, and has published a book, *Race, Evolution, and Behavior: A Life History Perspective* (Transaction Publishers, 1995), offering a sociobiological explanation for IQ differences among the races.

Perhaps this is the place to pause and discuss sociobiology as an account of the evolution of social behavior and its maintenance in a range of contemporary animal species, including humans. The term "sociobiology" was invented by E. O. Wilson, a professor of biology at Harvard and a world-famous expert on ants, to describe and explain the evolution of cooperative social behavior so characteristic of ant species as well as other social insects, such as bees. Wilson was struck by the fact that although ants and bees, on the one hand, and humans, on the other, are far apart on the evolutionary tree, they all share the trait of intergroup cooperation. Furthermore, he noted that the evolution of insect sociality was clearly the result of a particular kind of selection in which extreme cooperation within and among the different "castes" of these species was fostered by close genetic resemblance, cloning, or near-cloning from a single fertile female within the nest. Although he was well aware that the mechanisms of reproduction were not the same in social insects and primates (including humans), Wilson felt that genetic selection, including selection as an ongoing process in contemporary human groups, could be shown to continually reinforce social behavior. Wilson was also aware that social behavior often demands altruism, in which an organism sacrifices its life for the good of the group, and that, at least at first glance, such behavior is paradoxical. The latter because it contradicts the law of natural selection that says individual organisms that live to reproduce are the ones that pass their genes on to the next generation. This paradox was solved when it was noted that any individual saved as a result of an altruistic act by *any* relative would still carry at least some of the genes of the sacrificing individual, including the genes for personal sacrifice. This process, which takes the name "kin-selection," would thus encourage the selection for altruistic behavior throughout the group. Kin-selection, along with other suggested mechanisms that we need not discuss here, became the foundation for the general theory of social behavior that came to be called "sociobiology" and later "evolutionary psychology." For some psychologists and biological anthropologists, sociobiological principles were used as a direct challenge to the prevailing anthropological paradigm of *cultural* determinism (the idea that culturally learned patters of behavior best explain most of contemporary human behavior). Cultural determinism was first proposed by Franz Boas, the founder of modern American anthropology. Boas, trained as he was as a scientist, was fully aware that all human beings were the result of the long process of biological evolution, which gave rise to our species with a set of universally shared patterns of behavior, but throughout his life he was equally insistent that *group differences* in behavior found among contemporary human populations were due exclusively to culture and were based on learned, rather than biologically inherited, patterns.

This is where Rushton comes in, since, in addition to claiming a difference in IQ among the races, he proposed a sociobiological argument to explain these differences. Rushton's basic theory is that different reproductive strategies, under positive selective pressures, led to differences in intelligence among "the three races, Oriental, Whites, and Blacks." This will be discussed at greater length below, but simply put, in his 1995 book, *Race, Evolution, and Behavior,* Rushton suggests that Africans were selected for high fertility and a reproductive strategy that favored large numbers of offspring while Europeans were selected for low fertility that favored fewer, better-cared-for offspring with intelligence (high IQ) as the premium. In 1999, and again in 2000, an abridged version of this book was published and sent unsolicited, free of charge, to 35,000 psychologists, sociologists, and anthropologists in the United States and Canada. The reader can get a good idea of Rushton's ideas by consulting either of the abridged editions. According to him the socially negative effects of selection in blacks include: hyper-sexuality, low IQ, high crime rates, poor parenting, hyper-sexuality, and a high frequency of AIDS. (For crime and parenting see chapter two, for AIDS see chapter three, for intelligence see chapter four.)

High praise for the abridged version of *Race, Evolution, and Behavior* can be found on the inside cover of that book. Who are the enthusiastic supporters of this work? Rushton, I remind the reader, was one of the top recipients of grants from the Pioneer Fund. The same goes for his coterie of admirers. The names Michael Levin and Arthur Jensen will already be familiar to readers of this book. This chapter ends with a look at a third admirer, Hans J. Eysenck, one of Rushton's former professors at the University of London. Eysenck himself studied with Cyril Burt. Another promoter of the book, Richard Lynn, who said that Rushton deserved a Nobel Prize, has benefited from the largess of the Pioneer Fund, as have other "sponsors," including Linda Gottfredson of the University of Delaware and Thomas Bouchard of the University of Minnesota. Still another is Glayde Whitney of Florida State University, who wrote the introduction to the autobiography of David Duke, former national director of the Ku Klux Klan.

To explain the evolution of high intelligence in whites and Asians, and low intelligence in blacks, as well as the social consequences of low and high sexual appetites, Rushton claims that migrations from Africa, where the human species originated, created severe problems for survival that led to larger brains in those who were to become whites and Asians. But changes in brain size created a drain on development that was responsible for reduced sexual drives. This change produced a new reproductive strategy (the K strategy) favoring a small number of births, with improved care for offspring and hence lower infant and child death rates.

Allow me a few quotes from the 2000 abridged edition of *Race, Evolution, and Behavior.* To explain the evolution of high intelligence in whites and low intelligence in blacks, as well as the social consequences of low and high sexual appetites, Rushton says the following:

> The more north the people went "Out of Africa," the harder it was to get food, gain shelter, make clothes, and raise children. So the groups that evolved into today's whites and Orientals needed larger brains, more family stability, and a longer life. But building a bigger brain takes time and energy during a person's development. So, these changes were balanced by slower rates of growth, lower levels of sex hormones, less aggression, and less sexual activity.
>
> Why? Because Africa, Europe, and Asia have very different climates and geographies that called for different skills, resource usage, and lifestyles. (Rushton 2000, 25)

And on page 87 we find the following, which somewhat contradicts the comparison between an ecologically benign Africa and the challenging climate of Europe. "Africa is warmer than the northern continents, but it is a less stable habitat. Droughts, storms, and disease from viruses, bacteria, and parasites cause high death rates. Survival in Africa means having many young (r-strategy). In the more stable environments of Europe and Asia, success comes from having fewer young, but caring for them very well (K-strategy)."

Aside from the contradiction noted Rushton's argument for geographic determinism is logical (if out of date) and, taken without recourse to data, might seduce people unfamiliar with the findings of modern DNA and population genetics. Rushton either ignores (or is ignorant of) consistent data showing greater genetic variation within populations than among populations, no matter what parts of the world are compared. It is this data that led population geneticists to question the concept of race as it applies to humans at least as early as the publication of W. C. Boyd's *Genetics and the Origin of Species* in 1950. Since that pioneer work a huge amount of data has accumulated to confirm the early doubts about race. These are presented at great length in the monumental compendium of Luigi Luca Cavalli-Sforza et al., *The History and Geography of Human Genes* (1994), and in Cavalli-Sforza's most recent work, *Genes, People and Languages,* published in the year 2000. The latter includes material drawn from archeology and linguistics as well as genetics. But the history of doubts about race as an important factor in behavioral differences among human groups was expressed as early as 1879 by Charles Darwin himself in his book *The Expression of Emotions in Man and Animals.* Rushton distorts Darwin's mature ideas about this subject when he cites Darwin's *The Descent of Man* (1871) without mentioning the major hypothesis of the later book, to wit, that the differences among the races as far as emotional expression is concerned are minor and superficial. The central argument of *The Expression of Emotions* was Darwin's support for the theory of monogenetic origin for the human species versus the then-popular idea that each "race" evolved on its own from different early hominids (polygenesis).

The last chapter of the abridged, *Race, Evolution, and Behavior* is in the form of questions and answers put to Rushton about the so-called r-K theory of racial evolution.

Among Rushton's many obsessions about racial differences are his conclusions concerning racial differences in sexual avidity, penis size, and erectile angle. These are further "correlated" with AIDS frequencies in the "three races" (Rushton 2000, 80).

Rushton's web page, to be found at "google.com," lists his research interests as follows: "Studying behavioral genetics and sociobiology led me to explore the dilemma of why everywhere in Nature, birds of a feather flock together, which include behavioral genetics and sociobiology." Rushton claims to have found that mate choice with preference within rather than across races is the rule, and that this rule is determined by genes. He goes on to say that his work on race is more controversial because his research findings show racial differences in brain size and intelligence, as well as such factors as social organization and respect for the law. To explain these differences Rushton says they are due to evolutionary influences on gene distribution.

These two separate lines of research share more than a single common thread. They both ignore the influence of culture in human behavior, and they point not only to a strong bias for genetic explanations but also to the *sub rosa* point that nature disfavors any kind of interbreeding, including among the so-called races.

A final word about Rushton's sociobiological theory concerning the evolutionary reasons for IQ differences between blacks and whites. Remember that Rushton tells us that the African strategy was to have a large number of children in order to end up with a few survivors while the European strategy was to have few children but to take good care of them, ensuring their survival. In order to invoke evolutionary theory in general he also suggests that because evolution is parsimonious, which it is, higher reproductive rates demanded bigger and better sexual organs that were traded in for smaller brains and lower intelligence. Hence, his interest is in penis length and erectile height in addition to brain size. But the hypothesis concerning family size in different races begs the question of how far back in history low family size among Europeans can be traced. If my great-grandfather's family on my mother's side is any indication of family size at the turn of the century, then Rushton's theory falls under the weight of the 12 children that my great-grandfather sired. My wife's grandmother had 10 children. In general we know that small families among people of European stock are quite recent and are due to a series of cultural factors rather than to any form of biological selection. These include better hygiene and medicine from the beginning of the twentieth century onward, the rise of effective birth control, the increasing frequency of women in the market place, women's liberation, and so on.

As for Rushton's research on brain size and intelligence, he has published no measurements of his own, relying only on the work of other researchers. Throughout his many publications he offers a large sample of data drawn from sources of varying reliability (see, for example, Rushton 1992a, 1992b). One large data set comes from a stratified random sample of 6,325 U.S. military personnel in which cranial capacity was correlated with sex, rank, and race. In this study estimates of brain size were extrapolated from head measurements taken from living individuals.

After adjustments for stature and weight, the data are interpreted to show that men on the average have larger and heavier brains than women. Additionally, individuals of higher socioeconomic status have larger and heavier brains than individuals of lower socioeconomic status. Finally, Rushton claims to have found significant brain size and weight differences among the three major races. Individuals of northeast Asian ancestry have larger and heavier brains than Caucasoids, and Caucasoids have larger and heavier brains than Negroids. He also breaks his data down by sex and by rank (enlisted personal versus officers). Rushton offers the following brain size averages: men, 1442 cc.; women, 1332 cc.; officers, 1393 cc.; enlisted personnel, 1375 cc.; Mongoloids, 1416 cc.; Caucasoids, 1380 cc.; and Negroids, 1359 cc. Another data set was drawn from the International Labour Office in Geneva. It consisted of external head measurements gathered over a thirty-year period from men and women between twenty-five and forty-five years of age. After the usual adjustments for size and weight Rushton estimates from the data that Mongoloids have an average cranial capacity of 1312 cc., Caucasoids 1284 cc., and Negroids 1228 cc.

While the two studies show similar correlations between race and head size, the statistics offered are considerably different. The estimated brain size for Mongoloids in the two studies differs by 104 cc., for Caucasians by 96 cc., and for Negroids by 131 cc. These disparities are larger than the differences in brain size claimed for the races in either study taken alone. The first study shows a difference of 36 cc. between Mongoloids and Caucasians and 21 cc. between Caucasians and blacks, while the second shows the latter differences to be 28 cc. and 56 cc. respectively.

Rushton argues that because the brain is a very expensive organ (it uses about 20 percent of the body's metabolic rate in humans) increasing size must have had a strong positive evolutionary stimulus related to increasing intelligence. While this is certainly the case in mammalian evolution in general and primate evolution in particular, the argument ignores equally important changes in cerebral organization that occurred during the emergence of the primate line, particularly among hominids. It also ignores the fact that *normal* brain size in humans varies widely. Within the normal range very intelligent individuals have been found with relatively small brains and rather dull individuals have been found with relatively large brains. Anatol France provides the famous case for high intelligence in a small brain. His brain cavity was measured after death and his cranial capacity was found to be 1000 cc. at the very low end of the normal range.

More important, it is not accurate to argue for brain size correlations with intelligence within subgroups of a species from data on *between*-species evolution. The general trend in human evolution from the early *Australopithecines* through the *erectus* forms to modern *Homo sapiens has* been one of increasing brain size, but, as I have just pointed out, humans have a normal range with rather wide variation. Rushton's argument about brain size and IQ is based partially on a false analogy between primate evolution in general and living humans in particular. Furthermore, Rushton's claims concerning consistent brain size differences between sexes and among the races are taken from skull measurements on living subjects. These are highly suspect and can be compared neither to measurements taken of brains themselves nor of endocasts drawn from the interior of skulls after death. Skull thickness varies independently of brain size from individual to individual. Rushton is not unaware of this problem and attempts to deal with it by citing one study of living brains that uses magnetic resonance imaging. This study was, however, limited to a white sample of 40 college students. It showed a *small* correlation between measured brain size and IQ but ignored possible environmental factors and, more importantly, included no *intrafamily* comparisons.

Here I am reminded of those dull and bright rats described earlier. Let us return for a moment to these studies. Further experiments designed to investigate environmental influences on pure strain animals led to the following results: $S_{3\ (\text{“dull”})}$ rats were divided into two groups. One group was raised in an "enriched" environment that consisted of a transparent cage open to the visual stimulation of changing light. The cages were also equipped with running wheels for exercise and the animals were regularly handled by experimenters. The second group was raised in a "deprived" environment. These animals were isolated in translucent cages, had no access to exercise wheels and were never handled. When tested for maze running ability the rats raised in the enriched environment did significantly better than the rats raised in the deprived environment. Significantly, the better-performing rats had bigger and heavier brains than their deprived siblings. In addition, microscopic examination of nerve networks showed a considerably higher density of dendritic branching (nerve connections) in rats raised in the enriched environment when compared to those raised under deprived conditions (Ralph Holloway, personal communication, n.d.).

These results should surprise no one. While it may be true that individuals are born with all the neurons they will ever have, the development of the brain involves more than mere neuron number. Another major factor in brain function concerns the number of connections among neurons. It is now quite clear that the active development of the neural network is heavily influenced by early environmental stimulation.

The biologist Gerald M. Edelman has developed a complex theory of consciousness that takes this interaction between brain and environment into account. Edelman has published two books on the subject (Edelman 1987). His theories point to a strong environmental contribution to brain development *and* intelligence. This is consistent with the much earlier data on rats provided by Tryon (1940) and later followers.

Let us now turn to a detailed look at one of Rushton's major articles on brain size and race, "Cranial Capacity Related to Sex, Rank, and Race in a Stratified Sample of 6,325 U.S. Military Personnel," published in *Intelligence,* vol. 16, 1992.

Before treating the specific data from the military, Rushton discusses correlations between head size and socioeconomic status from a data set containing approximately 10,000 white and 12,000 black four year olds. This study showed "a small but significant correlation between head circumference and SES [socio-economic status] of origin within both populations . . ." (Rushton 1991, 402). Rushton argues that studies of brain weight at autopsy and endocranial (within the skull) measurements confirm the general findings of external skull measurement. He states that in the first study whites had a combined adjusted mean of 1323 cc. and blacks 1223 cc. The second study involved endocranial volume in a world sample of 20,000 crania controlled for region and showed Asian skulls averaging 1380 cc., European skulls averaging 1362 cc., and African skulls averaging 1276 cc.

Rushton Cautions: "These recent data on race differences do not go undisputed. Concerns range from the representativeness of the samples to the Appropriateness of the controls for body size. . . . Even the critics, however, acknowledge that '*some*' of the data are 'trustworthy' and in the direction claimed" (Rushton 1992b, 403; italics mine). What Rushton does not consider here is the possible environmental effect on whatever data in these studies was "trustworthy."

In his conclusions, Rushton, like so many of his colleagues plagued by their compulsive interest in racial differences, admits by way of disclaimer:

> It must be emphasized, however, that there is enormous overlap in most distributions. . . . For example, because race is only a weak predictor of cranial capacity (a 4 percent difference between the Mongoloid and Negroid average in this study [less for Caucasoids compared to Negroids, but this Rushton does not say] and head size is a weak predictor of intelligence (r=.30) it is clearly problematic to generalize from a racial group average to any particular individual. However, because there is about a 30 percent correlation between head size and intelligence test scores, these systematic and possibly causal relationships *are of great scientific interest.* (Ibid., 411, italics mine)

I would add that these *possibly* causal relationships are of great scientific interest only to a small coterie of researchers who grasp at straws convinced, in spite of all evidence to the contrary, that certain races are superior to others.

Let me now turn to the polemic, aired in the *Anthropology Newsletter,* an official publication of the American Anthropological Association, now called the *Anthropology News.* This took place between the Pioneer Fund, Transaction Publishers, and Rushton, on the one hand, and the majority of anthropologists, on the other, concerning the refusal of the editors of the *American Anthropologist,* also an official publication of the American Anthropological Association, to run an ad for the book in its pages. Robert Sussman wrote in response to complaints concerning the rejection of the ad by various professional journals, as well as its acceptance in some. Sussman, the editor-in-chief of *American Anthropologist,* responded in the *News* for May 1998 as follows:

> I understand that Transaction Publishers is attempting to get advertisements published for J. Philippe Rushton's *Race, Evolution and Behavior* in a number of major journals and has been successful in at least some of

> those journals published by Wiley. This is because the editors did not see these ads before they appeared in the journal. This is an insidious attempt to legitimize Rushton's racist propaganda and is tantamount to publishing ads for white supremacy and the neo-Nazi party. If you have any question about the validity of the "science" of Rushton's trash you should read any one of his articles and the many rebuttals by ashamed scientists.
>
> I am strongly against any publication of this ad in the *American Anthropologist* or any other journal published by the AAA. This would be inappropriate and offensive to the association members in general and to me. In fact, I feel so strongly about this issue that I would be willing to resign my editorship if this ad appeared in any journal that I was associated with.

In the same issue we find a letter from the president of the Pioneer Fund, Harry F. Weyher, denying that his foundation is racist and that grantees are free to publish any results they wish. What this letter fails to note is that whatever the Fund demands or does not demand of its grantees, all the ones I am aware of show the Foundation supporting research on race by individuals whose "results" can, in my opinion, be predicted in advance. The selective process is completely in the Foundation's hands.

Finally the same *News* prints a letter from Professor Jane H. Hill, president of the American Anthropological Association. Hill, responding to Weyher, says of the rejection:

> Personally, I find it astonishing that a social science house with Transaction's distinguished history would publish such a work. I see no benefit to the AAA in further dignifying it by advertising it in our journal. I take full responsibility for this decision . . .
>
> It is not the policy for the American Anthropological Association to censure minority views concerning professional issues. Articles to its journals are, however, submitted on a regular basis for peer review to well-known professional anthropologists. The editors of the various journals of the association do feel free to reject advertisements that they judge contrary to the standards of the association. Conflicting views that touch on sensitive political issues that go beyond mere *scientific* judgment can be found in the letters-to-the-editor column of the *Anthropology Newsletter*, including the Pioneer Fund, Transaction Publishers, and Rushton himself. Rushton's

> response, for example, to the American Anthropological Association's publication on race ("The AAA Statement on Race") was published in the *Newsletter* in the December 1998 issue.

At this point many of my readers may be scratching their heads and asking: Why bother to write a book criticizing ideas that are so clearly false and the expression of a tight minority of individuals? The unfortunate answer is that in the field of psychology, as opposed to anthropology, there apparently are many who swallow the IQ argument whole. The following letter, published in the *Anthropology News* for September 1998, from Jefferson M. Fish, a psychologist and professor at St. John's University in New York City, clearly illustrates his view of the danger of letting racist ideas professed by members of the academic community go unchallenged.

> As a psychologist member of the AAA, I have been reassured by its handling of matter of "race." . . .
>
> Anthropologists might be interested in a contrasting experience from the *APA Monitor,* the newsletter of the American Psychological Association. Its August 1997 edition carried a staff writer's front page story "When research is swept under the rug: Some of the best psychological research suffers for the sake of 'political correctness.'" This story portrayed Rushton (along with Arthur Jensen and others) as a serious scientist who has been victimized because his views are not "politically correct." As a psychologist informant, I can confirm that the story accurately reflects the predominant view in academic psychology. . . .
>
> Meanwhile the *Monitor* has turned down my request that they print the AAA's 1994 resolution on "Race" and Intelligence, the AAA's 1995 Statement on the Misuse of "Scientific Findings" to promote Bigotry and Racial and Ethnic Hatred and Discrimination, and the AAPA's [American Association of Physical Anthropologists] 1996 Statement on Biological Aspects of Race. I had suggested that, since APA is also in Washington, they cover the 1997 meeting—organized around the theme of race—and do a story contrasting anthropologists' understanding of race with that of psychologists. They chose not to because the *Monitor* covers psychology not anthropology.

Professor Fish extended his arguments concerning psychology's blindness to the false nature of race as a biological concept in an article published in the *American Anthropologist.* In this article, Professor Fish said the following:

> Psychologists' ethnocentric assumption that "emic" [culture based] categories of "race" in the United States are biological realities appears particularly intractable, . . . My frustration at the imperviousness of American psychology, as a cultural group, to attempts to challenge this assumption with scientific data from anthropology has led me to a perverse respect for the power of ethnocentrism. It has also led me to write about "race"—including efforts like this article aimed at encouraging anthropologists to help out. . . . (Fish 2000, 558)

Anthropology has been the only social science capable of dealing in a professional way with racist arguments that reflect pseudoscientific reasoning and "research" because the field, in this country at least, has, until very recently, united scientists in four subfields of the discipline: cultural anthropology, physical (biological) anthropology, linguistics, and archeology. Unfortunately, in this postmodern era, the profession is beginning to lose its focus in this respect as more and more departments of anthropology, which have until recently followed the traditional four-field approach, fragment into separate departments divorcing physical anthropology and its special competence in the area of race and evolution from the other three fields, which deal with various aspects of culture. The result has been that graduate students are no longer required to take courses in the "four fields." Archeology and linguistics, although they may remain under a single departmental umbrella with cultural anthropology, are often left by the wayside in both course requirements and comprehensive examinations. This has recently become the case in my own department, which even offers undergraduate students two possible majors, one in the four fields and the other exclusively within cultural anthropology. I am pleased to say that my archeological and physical anthropological colleagues still require their undergraduate students to take courses in cultural anthropology. This trend away from the four fields is rapidly leading to a new generation of cultural anthropologists unprepared to respond to racist nonsense. I find this situation most unfortunate. To give an example of the usefulness of the four-field approach I will cite yet another article from the newsletter of the American Anthropological Association, this one from the February 2000 issue. In this article, "Teaching Critical Evaluation of Rushton," the authors, all cultural anthropologists trained in the four fields, demonstrate a method for inoculating students from swallowing the views expressed in Rushton's *Race, Evolution, and Behavior.*

> We present one method we believe is particularly useful to help students develop a critical perspective on Rushton's work—having the students evaluate predicted associations between race and behavior using cross-cultural data. Few of the relationships predicted by Rushton find support when empirically tested and students are often annoyed to find that Rushton presents such relationships as if they had strong empirical support. Their annoyance often leads students to critically examine other aspects of Rushton's work, questioning his assumptions, logic, and data. For instructors this is an attractive situation since you don't have to take on a "moralistic" stance or present formal arguments against the work—you simply let Rushton's flawed ideas undermine themselves. (29)

Let me now turn to the last academic featured in this work, Hans J. Eysenck. His major contribution to the IQ argument was published in England in 1971 under the title *Race, Intelligence, and Education* and in the United States as *The IQ Argument.* In an article, "Science, Racism, and Sexism," Eysenck tells us that this book was written:

> . . . because of the considerable uproar caused by the publication in 1969, of an invited article by Arthur Jensen in the *Harvard Educational Review,* . . . Jensen has described in detail the persecution he suffered as a result of his scholarly and fully documented article. He and his family received threats that bombs would be planted in their house; he was personally attacked, his lectures broken up, his invited contributions to scientific conferences shouted down, reviewers misrepresented what he had said, lied about the facts, and made him out a racist and a fascist. . . .
>
> I wrote my book in order to introduce some sanity into what had become a political, ideological debate. All I did was to collect the relevant facts, and put them together, leaving it to the reader to judge. (Eysenck 1981, 217)

This quote appears in an article written as a defense against *all* those who dispute the IQ argument. It cannot be denied that there were indeed some violent reactions to the Jensen Report, but the majority of the published criticisms were dignified and scholarly in tone. One cannot say the same for Eysenck's response to critics of his own book.

Let me make one thing clear. I personally deplore any attempts to limit the academic freedom of *any* of my colleagues and am also an absolutist when it comes to First Amendment rights. BUT Jensen and his followers, not the least of them Eysenck, have used the unreasoned actions of a few as a smoke screen to defend their own position. This they have done not by reasoned arguments but rather by using what I have called a cleverly mounted reverse *ad hominem* in which *all* their adversaries are put into the same basket of unsavory crabs.

In the fall of 1981 I was present when Eysenck read a paper at a psychology congress in Lisbon. I remember arguing with Eysenck about IQ and race as did my French colleague Albert Jacquard. Eysenck presented a new claim at that congress. He was now able, he said, to measure IQ with a totally culture-free test. It could be read directly from EEGs (electroencephalograms). I have seen nothing in print since that time with which I can judge this claim. I can say, however, that even if we could measure IQ this way it would not demonstrate that the IQ so measured was under *genetic* control. Measuring IQ with a culture-free test (including brain waves) is not equivalent to determining what role heredity plays in its cause. Brain waves cannot answer the question of whether or not intelligence is wholly genetic, partially genetic, or not genetic at all.

Now let me return to Eysenck's article "Science, Racism, and Sexism." After bemoaning Jensen's fate Eysenck claims that his book met the same type of unreasoned attack, "When the book appeared the roof fell in" (Eysenck 1981, 218). Eysenck was hurt to the core by the book's reception and defended himself by claiming membership in the "old left," which had a strong dislike of racialism (note the term "racialism"), believed in equality of opportunity, and hated exploitation.

Eysenck says that when he began teaching about IQ in the English university system he followed the standard line that differences were due to environmental effects. He tells us that he soon began to see these arguments as flawed, evidenced by a growing body of data:

> If you matched black and white children in America, with respect to schooling, housing, and parental status and income, this hardly reduced the 15 point difference in IQ between the races; it came out at 12 points. Worse, if you took the children of black middle-class parents, coming from good schools and living in good surroundings, and compared them with the children of white working-class children, coming from poor slum schools and living in poor surroundings, still the white children came out better on IQ tests. (Ibid., 219)

The reader should note two things about this quote. First, the IQ difference attributed by Eysenck to genetics is 12 points, a figure higher than Jensen's by 4.5 points and that allows for almost no environmental effect on observed variation. Second, there is no way of checking his conclusion since he cites no published work to support his claim. Instead, he goes on to say that his new ideas were confirmed by the publication of Audrey Shuey's "great" book, *The Testing of Negro Intelligence,* which appeared in 1966. This book is a compendium of IQ and race studies, all showing an IQ deficit in blacks when compared to whites. What it is *not* is a critical analysis of these studies. It offers no insights into the problems raised by critics of IQ and race studies.

Eysenck follows his support for Shuey's book with a disclaimer that is, itself, curiously disclaimed:

> I re-read the whole set of articles relating to this problem, and emerged with the firm impression that Shuey was right. I also decided that having set my mind at rest, I would not myself publish anything on this problem—the blacks, or so it seemed to me, were having enough problems without me adding another one! But this decision was short lived. The publication of Arthur Jensen's monograph in the *Harvard Educational Review* brought the discussion to a boil, and clearly nothing could put the genie back in the bottle. (Ibid., 220)

Of course, Eysenck *could* have helped to put the genie back in the bottle by carefully analyzing the serious flaws in the Jensen Report. Perhaps he was impressed by the notoriety the article brought Jensen, his former student. As we have seen the intense press reaction to Jensen was generally as positive as it was careless. Eysenck's petulance may have been stimulated in part because his own book on race and IQ did not get the same attention in the American press, although it was widely reviewed in Great Britain.

Eysenck claims his book was written to raise a problem and, therefore, to help solve it. This problem was not one of race but one of a potential threat to the entire American population. Blacks, he tells us categorically, have low IQs but, because there are more whites than blacks in the United States there must also be more stupid whites than stupid blacks, and stupidity is the real problem in a society of growing technological complexity. Since IQ is genetic the problems caused by low IQ cannot be solved by traditional left-wing methods, since these stress environmental solutions such as the improvement of schools and educational opportunities.

Eysenck's reverse *ad hominem* dominates the next section of his article. There he categorizes four types of responses to his book. The first, he says, was a clear, positive, and thoughtful presentation of his arguments. However, he complains, although it was discussed hundreds of times, only a single "proper" review appeared. This was in the *Times* of London. The second and largest response to his book, he claims, was that which trivialized its contents.

The third category of review, the one Eysenck considers most serious and that appeared in the majority of newspapers, simply ignored his *IMPORTANT* work: "The Germans call this treatment *totgeschwiegen*—killed by silence—and it is a clever and very efficient method of dealing with an awkward problem. This method was adopted almost universally by the American papers" (ibid., 222).

The fourth category of reviewers were, in Eysenck's terms, "hatchet men." These hatchet men engage in five types of ploys as they set out to destroy a book. First, they criticize a book "without ever mentioning the facts and arguments stated by the writer to be the most important and convincing" (ibid., 224). Second, they quote authority, or, as he puts it, they "misquote" authority. Here Eysenck presents evidence that although critics said that professor Donald Hebb disagreed with his major argument, this authority actually never "denied the great importance of the genetic contribution to *individual* differences in intelligence; as I point out in my book, this is the crucial difference" (ibid., 224, italics mine). Here I expect that the reader will have noticed Professor Eysenck's own ploy. To state that genetics plays an important part (not the only part, by the way) in individual differences in intelligence is to make no claims about group (racial) differences. The major point of Eysenck's book is *not* that *individual* differences in IQ are primarily genetic but that racial differences in IQ are! Third, critics contradict arguments the author says he never made and disregard his own criticisms of these arguments. Here Eysenck offers no examples so I cannot respond. Fourth, he complains that some critics use "time distortion" to give an impression of inaccuracy in his work.

The fifth and last ploy cited by Eysenck is the attribution of opinions to him that are exactly the opposite of his real ones. Thus, he claims, the *Sunday Times* misrepresented him when they said he indiscriminately attacked attempts to improve Negro education because blacks were too stupid to benefit from them. Here Eysenck *has* a point. Nonetheless, it is fair to say that the main thrust of his book is not a plea for educational improvement. Rather, it is a clear attempt to prove genetic inferiority for blacks, at least as far as IQ is concerned. This is a blatant case of the tail wagging the dog in an attempt to deflect criticism.

After criticizing three unnamed London journalists, Eysenck proceeds to note with pride a review by the "very eminent geneticist, Professor C.D. Darlington." "Here, one might think, technical criticisms would come quick and fast; no such thing. The expert approved; it was the journalists who did not. Thus we seem to live in a very topsy turvy world, where expositions of scientific facts are criticized by journalists as not being factual, while the experts have no such complaints" (ibid., 225). Eysenck leaves out two very important facts: C. D. Darlington was in no way an expert on the genetics of race nor was he a competent authority on IQ. His contribution to genetic theory, which is by no means insubstantial, was in the area of the adaptive significance of sexual reproduction. More important is the fact that Professor Darlington was among the small group of English geneticists who were also racists. Of course he wrote approvingly of Eysenck's work.

When professionals *are* critical of Eysenck they are referred to pejoratively as being well known for their leftist opinions. In discussing a debate he had on British television, Eysenck remarks, for example, "the economist-educationist again was a leftist pundit, with no knowledge of either genetics or psychology" (ibid., 226). He goes on to say, "Then it was the turn of the two colored students who had been elected to represent the interrupters. What they said was, unfortunately, almost pure nonsense" (ibid., 228). Unfortunately, we are not told what they said.

Most of the article I have been examining here deals with public reaction to Eysenck in various contexts, journalistic reviews of his work, television appearances, and university lectures. Having been present in the late 1960s and early 1970s at meetings where ideas unpleasant to the left were aired, I will not claim that Eysenck has lied about the tone of some of these encounters. I myself have been present when scholars who believe in genetic arguments concerning human behavior have been treated with more than simple discourtesy. In 1978 I was chairing a meeting of the American Association for the Advancement of Science when a group of leftists opposed to sociobiology disrupted the meeting by pouring water over the head of E. O. Wilson, who was present to debate the issues around this subdiscipline of biology. Therefore, let me repeat: I condemn any such behavior uncategorically. I, too, have experienced violent and unthinking reactions from fringe people on the right who have attacked *my* ideas on race. In addition, over the years I have been the recipient of a good deal of anonymous hate mail, precisely because I have written against the work of Jensen, Eysenck, and others. But the actions of a small minority of intolerant individuals should not be used to cloud the issue under discussion. There is, of course, another lesson here as well. The work of scientists should be judged on its scientific merits. The thoughtless acts of a small minority should not be used to turn individuals who do poor work into martyrs.

Whether or not Eysenck was an "old leftist" in his younger days, he certainly turned full circle in thinking. Thus, when he makes a plea for tolerance it is a plea for his own cause that is phrased in the most cautious of terms. He says finally of the race and IQ issue: "Can we attribute these to genetic or environmental causes? Can we do anything about reducing or abolishing them? No one knows for sure, and our best way is obviously that of critical and cautious experimentation. My book constituted an attempt to survey what was factually known about such problems, and to suggest what conclusions, if any, could be drawn from the facts" (ibid., 236).

Yet this is a man who could say later in the same article that: "The evil consequences of ignoring scientific facts, and believing instead ideological preconceptions, are well illustrated by the American 'busing laws,' enforcing racial integration by busing white children to predominately black schools. These laws spawned by unscientific thinking and willful ignorance, have had predictable effects . . ." (ibid., 245).

I hope to have already made it clear that I support the research of legitimate scientists into whatever field interests them. All scientists, however, must realize that they have a strong obligation to the public at large. Great care is needed to verify data and theories when they may effect public policy. They should in no way be immune to scrutiny from their colleagues, who have an obligation to report failures of research design and interpretations of results.

In my opinion Eysenck is most disingenuous when at the conclusion of his article he says: "A good heart is not enough when it comes to designing effective social action to help blacks, or women, or any other group that may be having difficulties; a good head, free from ideological preconceptions, is also required if the action taken is not to have effects directly counter to the intentions of its author" (ibid., 247).

I believe that the late, great population biologist and evolutionist Theodosius Dobzhansky should have the final word on this particular issue. To paraphrase his words spoken at a public meeting many years ago: The only experiment that will provide the definitive solution to the race and IQ question will be to create a society of truly equal opportunity in which every individual will be allowed without the constraints of poverty and race prejudice to reach their own potential. These are the conditions that will allow whatever genes are there to penetrate to their full 100 percent. I will only add that this is the American ideal.

CHAPTER TEN

EPILOGUE

Other Racisms and Related Matters

In this chapter, after a last bizarre excursion into reductionist behavioral genetics, I turn to problems that go beyond biology and IQ to other forms of racism. The examples to come are drawn from my own experience as a teacher and researcher in the field of cultural, rather than biological, anthropology.

PSEUDOSCIENCE = BAD SCIENCE

On July 13, 2000, *Avui,* the Catalan language daily of Barcelona, published an interview with Professor Victor Faris, professor of history and philosophy at the Free University of Berlin. Among the topics covered was his discovery of documents proving close cooperation between Hitler's Third Reich and Franco's Spain during World War II. These concerned "research" pursued by Spanish scientists beginning in 1939 and using subjects in Franco's concentration camps, where former soldiers in the Republican army and other anti-Franco Spaniards were incarcerated. Professor Faris's data was drawn from the archives of the Center for the Study of Criminal Medicine, which documented the work of Spanish doctors, psychologists, and psychiatrists working in cooperation with

Nazi scientists. In the interview the professor cites material to show that Franco's "experts" paid special attention to the brains of deceased Catalans and Basques, two peoples who provided the fiercest resistance to Franco's nationalists during the civil war. The initial study was designed to look for evidence of an inborn tendency toward violence, this from adherents to a movement that had as one of its battle cries "Long Live Death!" The longer-range goal was to discover whether certain individuals had an inborn proclivity for Marxism that could be linked to a hypothetical "Marxist" gene. Biological determinism indeed!

FOLK SCIENCE = BAD SCIENCE: HOW TO MISUSE RELIGION TO HIDE RACISM

In 1943, in the midst of the World War II, the U.S. Public Affairs Committee published a pamphlet entitled *The Races of Mankind,* authored by two Columbia University anthropologists, Ruth Benedict and Gene Weltfish. The pamphlet was written under the supervision of several well-known Columbia professors, including L. C. Dunn, geneticist; Otto Klineberg, social psychologist; and Dr. Marion Smith, anthropologist. Although somewhat anachronistic in light of today's knowledge contradicting the *biological* validity of race, the pamphlet, written in simple language for the lay public, reviewed the then-current evidence for the essential oneness and equality of all the varieties of the human species.

The Races of Mankind begins:

> Thirty-four nations are now united in a common cause—victory over Axis aggression, the military destruction of fascism. These United Nations include the most different types of men, the most unlike beliefs, the most varied ways of life. White men, yellow men, black men, and the so called "red men" of America, peoples of the East and West, of the tropics and the Arctic, are fighting together against one enemy. . . .
>
> We need the scientist just as much on the race front. Scientists have studied race. Historians have studied the history of all nations and peoples. Sociologists have studied the way in which peoples band together. Biologists have studied intelligence among different races. All that scien-

> tists have learned is important to us at this crucial moment of history. . . . This booklet cannot tell you all that science has learned about the races of mankind, but it states facts that have been learned and verified. We need them. (3)

The Races of Man goes on to provide short and easily read sections concerning such topics as blood, skin color, head shape, race mixture, language as an aspect of culture rather than biology, culture in general as learned rather than inborn, character as a product of culture, civilization as cultural rather than biological, and the future of race prejudice. It describes (with some exaggeration) how much organized labor, churches, and the government all contribute to fighting racism in the United States and ends by noting the challenge that racism poses for democracy in the context of the war:

> With America's great tradition of democracy, the United States should clean up its own house and get ready for a better twenty-first century. Then it could stand unashamed before the Nazis and condemn, without confusion, their doctrines of a Master Race. Then it could put its hand to the building of the United Nations, sure of its support from all the yellow and the black races where the war is being fought, sure that victory in this war will be in the name, not of one race or of another, but of the universal Human Race. (31)

Early in the pamphlet, Benedict and Weltfish (perhaps to gain support for their position on race from a public ill informed about science) call upon the Bible to bolster their arguments concerning racial equality:

> The Bible story of Adam and Eve, father and mother of the whole human race, told centuries ago the same truth that science has shown today: that all people of the earth are a single family and have a common origin. Science describes the intricate make-up of the human body: all its different organs cooperating in keeping us alive, its curious anatomy that couldn't possibly have "just happened" to be the same in all men if they did not have a common origin. . . .
>
> The races of mankind are what the Bible says they Are—brothers. In their bodies is the record of their Brotherhood. (5)

This section of *The Races of Mankind* is illustrated with a line drawing of the human family tree. It shows Adam and Eve in their garden complete with snake and apple. At the top of the branches of the Tree of Knowledge are four male figures: a black, a brown, a yellow, and a white. Adam is shown clothed only in his traditional fig leaf while Eve is partially clothed to hide her private parts. Innocent enough? Apparently not! *The Races of Mankind* was brought before an official congressional committee charged with investigating its fitness for distribution to members of the armed forces. The committee's majority was composed of conservative (Democratic) representatives, elected more or less exclusively by their white co-residents in the one-party South. (At the time the region was segregated, and few African Americans had won the right to vote.) Outraged by the message portrayed in *The Races of Mankind,* these representatives of "the people" needed some excuse to refuse public funds for the distribution the pamphlet. They found it in the illustration. What was the problem in the eyes of these officials? Nothing simpler! Both Adam and Eve in the drawing were taken as a denial of the Bible's teaching. Why? Because both figures are drawn in possession of bellybuttons, which, according to Genesis, was impossible since Adam was created out of clay by God and Eve out of Adam's rib!

OTHER RACISMS = BLAMING CULTURE

Xenophobia, a common enough phenomenon, sees the superiority of one's own culture over all other cultures. While the reasoning behind xenophobia can, like racism, be based on false genetic arguments, it need not be. Religious differences are often the cause of hatred among groups, and even such simple cultural differences as eating habits can lead to the denigration of one culture by another. When religion and food are combined the result can be truly bizarre. Take, for example, eating pork. Those who refuse to eat it for religious reasons are suspicious of those who do, since pork is considered unclean and polluting. Those who favor pork in the diet are often just as suspicious of those who refuse to eat it. Pork is an important food item among Europeans, particularly among small-scale farmers, and in many cases is an essential part of the diet for the rural poor, who have little access to other forms of animal protein. Pigs are cheap, easy to raise in small numbers for domestic use, and can be fed primarily on remains from the kitchen table. They grow quickly, and practically every part of the animal can be used. Pig meat is also easy to preserve, salted, smoked, or smothered in lard. Furthermore, pig fat is frequently eaten by the poor as a substitute for butter. It should be no surprise that the Jewish and Muslim taboo on pork required an explanation, particularly for those who depended on the pig for survival and who additionally saw Jews as quintessential and dangerous outsiders. For example, in southern French rural areas a persistent folk tradition that continued well into the twentieth century was used to explain this taboo. Jews, the tradition said, refused to eat pork because they were descended from pigs and one does not eat one's ancestors!

That xenophobia need not be based on genetic arguments was shown recently in one of the two bloody wars in the former Yugoslavia. The three ethnic groups involved, Serbs, Bosnians, and Croats, not only share a common genetic heritage but they also speak the same language, known today as Serbo-Croat. The major difference among these people is religious. Most Serbs are Orthodox Christians, the Croats Roman Catholics, and the Bosnians Muslims. During the Bosnian War criminals among Serbian soldiers raped Bosnian women not merely as a terror tactic but specifically to create Serbian children, a fact that, of course, negates any genetic argument. Those born of these forced sexual encounters were taken from their mothers to be raised as Serbs—no better proof that genetics played no part in this particularly crude and cruel form of cultural domination.

BLAME IT ON THE POOR, BLAME IT ON THE MINORITY

Unfortunately, history, up to modern times, has seen many instances of racism and xenophobia, sometimes based on false genetic arguments, at other times on the ideology of religious and/or cultural superiority, often with deadly consequences. But racism and xenophobia in their many forms are not the only problem. Cultural anthropologists themselves have gone astray when they have been too quick to explain complicated social situations as the result of sub-cultural differences. This is particularly the case with the poor in the United States. In the early 1950s a small number of cultural anthropologists began to focus on the urban poor in the United States and Latin America. Cities in the United States, particularly on the East Coast, which was the traditional starting place for immigrants from abroad, provided a rich ground for research. In Latin America urban centers have been the targets of international migrations from poor rural areas. Indians and mestizo peasants with cultural backgrounds often different from those of city dwellers occupy the slums and squatter settlements of many Latin American cities. They form the bulk of the urban poor in these countries.

The best known of the early students of urban poverty is the anthropologist Oscar Lewis, who published a series of books on the subject, beginning with *Five Families* in 1969. These were based on taped interviews with a small number of informants. Lewis realized that this method violated the rule of statistical sampling but felt that what was lost in numbers could be gained from an in-depth analysis of the life histories of real people. In fact Lewis did produce a body of unique documents that were deeply moving. Instead of the cold statistics and abstract view of human life found in so many sociological and anthropological studies, Lewis presented his readers with individuals caught in the web of urban poverty. Lewis's original study focused on the Mexican family. This was later followed by an investigation of family life in the Puerto Rican slums and in New York City.

It was from this work that Lewis developed the concept of "the culture of poverty." His idea was that poor people (reacting to the social position in which they found themselves) develop a set of unique cultural characteristics that become *self-perpetuating*. If so, the poor become the victims not only of economics but of their own culture. Lewis himself described the culture of poverty as an adaptation and reaction of the poor to their marginal position in capitalist society. He characterized the poor as follows: They are fearful of society at large; they are hostile to political authority and hate the police; most of them are not members of an organized church and rarely participate in elections; and they lack a coherent social organization. The poor are said to live in the present. They take gratification wherever and whenever they can. Thus, among the poor saving is unlikely, as is placing a value on education as a way out of poverty. These negative values operate against the adoption of behaviors that might allow individuals to escape from a slum existence.

Lewis was a liberal who had a great deal of sympathy for, and empathy with, the poor people he studied. But the picture he drew of the "culture of poverty" was biased by his exclusive focus on the poor rather than a more global view of the culture, in which the poor only constitute a part. Additionally, some of the characteristics attributed by Lewis to the poor are by no means exclusive to the lower class. Hostility to political authority is not absent from the middle class. The poor's fear of the police is often justified since they are more likely than others to be the victims of police brutality. Blacks in the urban ghetto, for instance, are more likely than whites to be shot if they are seen running near the scene of a crime. In very recent years New York City has seen many blatant cases of police brutality perpetrated on African Americans and other minorities with no justification. While middle-class people can usually expect to receive bail and fair trials, the poor are often incarcerated for months before their cases come before judges. The lawyers assigned to defend the poor and racial minorities are often incompetent, as recent studies of individuals condemned to death for murder have shown. In the case of suspected lesser crimes the poor are often subject to much stiffer sentences than members of the middle class.

It is true that the poor live on the economic margins of society because they are unable to share in many of the economic benefits that are common in the middle class. It is also true that, almost by definition, many poor people do not save money, but this is due more to their precarious economic position than to their culture. The poor are generally excluded from the "credit card society" that is so characteristic of modern middle-class American culture. This means of living above one's head is the privilege of the affluent. There is an old joke about an American millionaire who came to the United States from Europe to make his fortune. "When I came here I was a failure," he is reported to have said; "I only had a dime. Now I am a success because I am ten million dollars in debt." There is more truth than meets the eye in this story. People who owe small amounts are often hounded by collection agencies. Large borrowers, certainly not members of the poorer classes, are often treated kindly by banks nervous about the recovery of their large investments.

Living in the present is frequently the only possibility for those who cannot accumulate a nest egg for the future. Nonetheless, there *is* evidence that the poor do plan and carry out concerted action that is future oriented. The anthropologist Anthony Leeds (1971) studied several squatter settlements in Latin America. These consist in the majority of cases of illegally constructed houses. The rule in many countries with a Latin heritage is that if a person can put a roof on a structure in the course of one night without attracting the attention of the police, the authorities have no right to evict the tenant or destroy the house. To build such a dwelling in so short a time, an individual must secretly accumulate the necessary building materials, muster a work force, and build in great haste. All of these actions take forethought and organization. Leeds also found that, as this type of settlement develops, a social structure evolves that is both unique and outside the usual pattern of legal authority. This social structure is often invisible to the middle class, including the middle-class social scientist.

The poor have many ways of dealing with society at large. These vary with circumstances and cultural background. My own first ethnographic research consisted of an extended study of an African American Pentecostal church. Although the members of this church were primarily from the urban poor, they exhibited few of the qualities described by Lewis as characterizing the culture of poverty. When this church is viewed in the context of American society at large its development can be seen as an organized if unconscious adaptive reaction to tough social conditions.

The United House of Prayer for All People on the Rock of the Apostolic Faith was founded in 1921 by C. E. (Sweet Daddy) Grace, a light-skinned Portuguese black immigrant from the Cape Verde Islands off the coast of West Africa. Grace settled in New Bedford, Massachusetts, a city with a large Portuguese immigrant population. The growth of the church in New England was slow, and Grace moved to the South, where he rapidly developed a substantial following. As his reputation as a miracle worker and curer grew, the sect spread back toward the North. This growth was also facilitated by the great black migrations to the urban centers of the Northeast that were taking place at the same time. In New York City Grace opened a storefront church in Brooklyn in 1930 and one in Manhattan in 1938. The first actual church building in New England during this period was built in 1956. After that congregations were established in Stamford, Hartford, and Bridgeport, Connecticut, all cities with substantial black populations. Sometime during this period Grace also established himself in Detroit, but the sect had little success in Chicago. A Los Angeles branch was also founded, but the base of the Church's membership consisted of people from the South and the Northeast. During the depression and World War II the Church grew into a large organization. At one point Grace claimed over one million members, but several thousand is a more likely guess. Compared to other, often ephemeral, storefront churches, however, Grace was certainly successful. His congregations were able to construct distinctive church buildings, and businesses associated with the sect grew into rather large enterprises.

During my study, in the early 1960s, the Church took no direct political stands, but Daddy Grace associated himself with the freeing of the slaves. The words of a frequently sung Church hymn, "Lincoln talked about it but Daddy made it true," reminded the faithful of Grace's direct intervention with God and the state for the benefit of African Americans. Grace capitalized on his light skin to present himself to the faithful as a benevolent white man who had come to bring equality to blacks. While politics were avoided, patriotic symbols were, and, since Grace's death in the 1960s, are still, used. For example, churches around the country are painted red, white, and blue as a sign of their status as an American church. Wealth, materialized in the form of dollar bills, is also a key symbol of the Church. Daddy Grace's visits to local parishes around the country were punctuated by gifts presented to him by the faithful. These consisted of elaborately constructed symbolic objects (for example, crosses or flower bowers) covered with real paper money. Additionally, during services followers would press money directly into Daddy's hands as he sat on the altar platform. He would give some of this money back to individuals as they came to the platform in salutation. This "redistribution" of wealth was taken as a sign of Grace's largess, as well as a manifestation of the financial power of the Church. Although in reality a good deal of the money collected during services found its way into the leader's pocket (over the years he accumulated a substantial fortune), a portion of it was returned to the faithful in the form of food served in Church restaurants, donated clothes, and subsidized rents in Church-owned apartment houses. While the poor who constituted the majority of the membership could do nothing with their individual meager funds, contributions, invested in a block by Grace, could be used for the benefit of all. On the symbolic level Grace's substantial personal possessions, vaunted to the membership rather than hidden from view, added to the glory of each Church member. On the social level the Church provided an alternative community to the faithful, who were expected to attend services on a nightly basis. Members occupied different church functions (ushers, members of the band, elders and deacons, the women's auxiliary, as well as workers in the many Church enterprises). Drinking, smoking, and drug use by the faithful was strictly forbidden.

After Grace's death, during the rise of black pride and black power, his successor used his own identity as an African American to shift the Church's emphasis toward the emerging political awareness in the black community, but the rest of its original symbolism has remained intact. The development and growth of this sect illustrates one way poor members of minority groups are able to organize themselves within their own communities and also in the face of the wider society. The United House of Prayer provides satisfaction to its members not only because it promises salvation (other churches make the same promises) but also because it objectively improves their economic status within the community.

Members' lives are enriched by the alternate social structure provided by the Church as well as by the material benefits that accrue to them. The decline in social pathology (alcoholism, drug addiction, and crime) is a further benefit that results from the solidarity provided by this community of the faithful.

While the United House of Prayer represents only a minority of individuals within a much larger urban poor minority population, it serves as an example of how churches within such communities have created alternative ways of dealing with ghetto problems. This example, which is only one of several, also shows that the poor—and probably any subculture imbedded in a large cultural system—must be analyzed in the context of wider social and class relationships. It also shows that although material forces play an important role in the development and orientation of institutions like the United House of Prayer, such evolving systems are also shaped by internal cultural elements. The Church was not just an adaptation of a segment of the poor to American society in some abstract sense. Instead it developed out of historical and cultural roots that existed as part of the rural black subculture.

ON THE EXOTIC STATUS OF THE "OTHER"

Within the completion of Western colonial empires at the end of the nineteenth and the beginning of the twentieth century, a new phenomenon emerged. This was the exhibition of the *other* in the context of circuses, zoos, and museums alongside exotic animals and other natural forms. This practice can still be seen today in the grouping of foreign ethnic exhibits, including dioramas complete with life-sized human figures dressed in native costume, in what are known as "natural history" museums. In the recent past such exhibitions, using live people, were integral parts of World's Fairs and other such settings, in which "native habitats" were constructed to be populated by indigenous people dressed in equally "native" costumes. These "specimens" were essentially frozen in time and presented as part of a vanishing and distant world as if their cultures were static and unaffected by the events that drive history for the rest of us.

The French monthly *Le Monde Diplomatique* for August 2000 carried a long article by Nicolas Bancel, Pascal Blanchard, and Sandrine Lemaire entitled "Ces Zoos Humains de la République Coloniale." In it the authors discuss the simultaneous emergence of "zoologic" spectacles, which involved the display of exotic peoples from what we now call "third-world" populations in several European countries. They reproduce a quote from one of these that lasted in Frankfort, Germany, from 1880 to 1914: "Australian cannibals, male and female. The one and only unique colony of this savage race, strange, disfigured and the most brutal ever taken from the interior of savage countries. The lowest order of humanity." (16, translation mine)

The authors go on to note that between 1877 and 1912 thirty highly successful exhibitions of this type were mounted in Paris at the Jardin Zoologique d'Acclimatation. I might add that the same success occurred at the turn of the century in New York's Bronx Zoo, where a live African

Pygmy was exhibited in a cage as part of the primate section. During the St. Louis World's Fair of 1912 a large number of exhibits focused on native peoples who were housed in copies of indigenous dwellings, dressed in tribal costumes, and who performed traditional crafts before the eyes of the public. It was at the fair that the first woman press photographer in the United States, Jessie Tarbox Beals, took a series of popular photos to be published at the time in several newspapers and magazines. Ironically, these same photos were published in the 1980s in the *New York Times Magazine* to illustrate native peoples in their own countries, the newspaper was apparently unaware that they were taken in the United States and dated from 1912!

The authors of the *Le Monde Diplomatique* article point out that this type of exhibit resulted from three related phenomena: the construction of an imaginary "other," the development of a belief in racial hierarchies, and the vaunting of colonial empires then in development. This historical reality combined with such exhibitions and gave rise to a vocabulary that stigmatized the savage as driven by blood lust, marked by an obscure fetishism, and stunted by beastly atavism. These ideas were reinforced by iconographic displays of "an unheard of violence" supporting the idea that such natives were subhuman, culturally stagnant, and living on the frontier between humanity and animality. Furthermore, the authors point out, these cultures, taken in the context of the now-outmoded idea of unilineal social evolution and situated at the bottom of that scale, *were* nonetheless civilizable and, *therefore, colonizable.* Thus, the Europeans involved in such an enterprise were engaged in a noble act!

While exhibits of this type would be unthinkable in the United States today, I did find a contemporary example in the course of my current research on Catalan identity. This was in an out-of-the-way natural history museum located in the small town of Banyoles, about thirty miles from the border between Spain and France. The museum holds the collection of Francesc Darder, whose career as a professional veterinarian and amateur taxidermist reached its zenith in the late nineteenth century. There, until very recently, in "The Hall of Man" the visitor could view a (literally) stuffed human figure: a man of the San People of Botswana dressed in a leather apron. The figure holds a spear on one hand and a leather shield in the other. Although the average San has light brown skin, the figure is painted black, perhaps to conform to the viewer's preconceptions concerning Africans in general.

The figure, finally removed from public view in 1998 under pressure from UNESCO, a group of African leaders, and the Spanish government, would not merit much attention if it were not for the long polemic it aroused in a wide segment of Catalan public opinion. Articles dealing with this issue in the Catalan language press, including letters to the editor, were in the vast majority against the removal of the exhibit and critical of the individual, Dr. Alphonse Arcelin, a medical doctor of Haitian origin (now a Spanish citizen) who was the first person to demand the exhibit's removal. (Dr. Arcelin had discovered the exhibit just before the Olympic games that took place in Barcelona in 1992.) As secretary of the socialist party of the coastal resort of Cambriles, where he practices, his first action concerning the "Black Man" was to ask his party to take a stand against the exhibit. When this was refused he attempted to put pressure on the autonomous Catalan government to act. According to Dr. Arcelin, whom I was able to interview in 1997, the president of the Generalitat (the Catalan government), Jordi Pujol, asked him to delay any action that might hurt the operation of the games until the event ended. Again, according to Dr. Arcelin, after the games all his requests for action were ignored with the excuse that the mayor of Banyoles was the only individual who had the authority to remove the offending exhibit. Dr. Arcelin then put pressure directly on the Banyoles mayor by bringing a lawsuit against him demanding damages (to be donated to charity) as well as the removal of the offending exhibit. In order to understand the press reactions that followed, the reader needs to know that Catalans, whose autonomy and right to speak and teach their own language was restored only in 1977, two years after the death of Franco, the dictator who ruled Spain from the end of the civil war in 1939 to 1975, have a brand of nationalism that is based more on language than on genetics. Additionally, Catalan nationalists tend to be hostile to criticism from outsiders. Dr. Arcelin, although fluent in Castelian as well as several other European languages, does not speak Catalan. (His medical education took place in Seville, he is married to a Sevillian woman, and most of his patients are foreign tourists.)

Let me now present excerpts from the Catalan press that illustrate the point I am making concerning the hostile, in some cases racist, and, at the very least, insensitive attitude that Dr. Arcelin's efforts aroused.

On March 4, 1997, the Girona newspaper *El Diari* (Banyoles is located near Girona) noted that the Spanish minister of foreign affairs had threatened a judicial action against Banyoles if the exhibit is not withdrawn from the museum. The article began: "Without having listened to the arguments of the City Council of Banyoles concerning the social and cultural value of the exhibition of the 'Black Man' the Spanish government has decided to avoid a deterioration in its relation with African States by calling for the withdrawal of the exhibit."

On March 6, 1997, *El Punt,* another Girona newspaper, published two short opinions in a section called "Applause and Boos." One of these notes that "What began as child's play and went on to become an international incident was about to end with the closure of the hall of man. Mayor Solana [the mayor of Banyoles] is to be congratulated for his patience during the polemic." The other opinion criticized Dr. Arcelin for his refusal to be satisfied by the closing of the Hall of Man as well as for his demand that the "Black Man" be sent back to his home country in Africa for a proper burial.

On the same day *El Punt* published a statement by the vice-secretary of the left nationalist Catalan party (the ERC) to the effect that Dr. Arcelin should learn to speak Catalan and thus show respect for the people and the nation of Catalonia. He was quoted as saying:

> The arguments used by Dr. Arcelin in favor of the removal of the "Black Man" and its subsequent burial or cremation should lead him to reflect on the character of the country that has welcomed him in its midst. His arguments in favor of respect for the human condition, and human rights, as well as respect for differences among peoples and cultures, are the same arguments that Dr. Arcelin should have considered in recognizing the special personality of the nation, the language, and the culture of Catalonia.
>
> Arcelin ought to defend, with the same emphasis, the special cultural values of Catalonia that he uses in the Darder affair. In respect to a minority culture taken in its international context like Catalonia, where the right to diversity is fundamental, Arcelin should be coherent and should speak in Catalan.

On March 11, 1997, *El Punt* informed its readers that:

> In an article published in *La Vanguardia* [another Catalan newspaper and the major Castelian language daily in Barcelona] the writer offers to have his body embalmed and exhibited in the same display case as the "Black Man." In so doing he hopes to denounce another form of racism. [On the basis of what he has heard about the issue] . . . which revolves around the exhibition of a person of the black race and has nothing to do with the exhibition of a human body.
>
> The Darder Museum exposed the "Black Man" as a mark of respect as well as scientific rigor to show the equality of all human races. . . . The "Black Man" of Darder is a brother because in life we are all equals. If the lawyer believed that in this exhibition there was even a minimum intention of racism he would not have made the offer.

On March 19, 1997, *El Punt* noted that the museum and its Hall of Man had been reopened subsequent to the removal of the "Black Man" to storage. The article also commented on a lecture given by Dr. Arcelin about the case during which he was criticized by a large segment of the audience. Some people went so far as to accuse Dr. Arcelin of hatred toward whites and severely criticized his attitude, which they said *threatened the integrity of Catalans!*

This removal of the offending exhibit was not the end of the affair. On the front page of *El Punt* on April 16, 1997, a headline read: "Com treure's el mort de sobre." ("How to get rid of the body.") The story notes:

> To bury the "Black Man" a special tomb will have to be created. If he is cremated the body will have to undergo a physical operation. Any repatriation will also be complicated because the body will have to be claimed by the country of origin. . . .
>
> In order to bury the body a specially large tomb will have to be created because of the position in which it is preserved. If cremation is the choice it will first be necessary to remove all noncombustible materials.

I will spare the reader further details of this article except to say that three-quarters of page three of the same issue are devoted to this "difficult" problem. What I cannot leave out is a smaller, separate article on the same page that bears the headline: "Internal organs removed, without bones and with his penis reinforced."

Up to this point no paper had published any sexual references to the "Black Man." Here I am tempted to see this example as a pendant for the Hotentot Venus, whose exposition live and later dead was the rage of Paris at the end of the nineteenth century. It was the putative sexual organs of the Hotentot Venus that brought her to the attention of the scientific community as well as of the public at large. After her death her private parts were dissected by no less a person than Broca (the discoverer of an area of the brain associated with one type of aphasia). Today these remains are preserved in a bottle placed gracefully out of the public eye on a shelf in the Musée de L'Homme along with her skeleton! Recently her descendants in South Africa demanded that her remains be deaccessioned and sent back to her homeland for burial. The French government agreed to do so in March 2002.

In a paper delivered to the women's study program at Columbia University in 1996, Yvette Christiance noted that the Venus's skin color was systematically treated as darker than it actually was by those interested in her physical anthropology. This subjective view of skin color provides yet another link between Venus and the "Black Man" of Banyoles. Lest one forget, let me remind the reader that the "Black Man"'s kidnappers, not satisfied with his light skin, painted it black to conform to the supposed African type specimen, and, probably, as a reinforcement of his "primitive" status.

The same April 16 article in *El Punt* went on to give details about how the "Black Man" probably died and about the probable disposal of all his internal organs and most of his skeleton, with the exception of the skull and extremities. It concludes by saying: "The body was probably stuffed with material of little density *with the exception of the penis which was reinforced with a stiffer material to better maintain the morphology of the organ*" (italics mine).

The April 16 *El Punt* also sheds light on the opinions of people in Banyoles and its environs. On page six the paper publishes a letter that begins: "Leaving aside the racist considerations of Dr. Arcelin, the inhibitions of the Generalitat, the precepitation of Mr. Solana, the pants lowering of Mr. Matutes [the Spanish foreign minister], and the deceitful actions of UNESCO, there are weighty reasons to conserve the Bushman of Banyoles. Arcelin has profited by the moment to insist, by reason of mania or persecution complex, that our friend of more than three generations be removed from his display case. . . . What would UNESCO's position have been if the 'Black Man' instead of being exhibited in a small, out of the way town, was the property of the British Museum or the Louvre?"

On April 22, 1997, an article in *El Punt* demonstrated to what lengths Mayor Solana was prepared to go to keep the "Black Man" (one way or another) in Banyoles. The headline reads: "A mold of the Bushman is made under the order of the mayor of Banyoles." The article goes on to say, "According to one of the three sculptors charged with the task, the mold was made of a special silicone and the carcass that covers the silicone is polyester. It was fantastic, the mold emerged in a state of perfection."

The April 29, 1997, *El Punt* published an op-ed by Joan Abril Espanol, a sarcastic review of the history of Dr. Arcelin's campaign to have the "Black Man" removed and sent back to Africa for burial. He ends by saying: "Why is it that the government of Senegal, the Minister of Foreign Affairs, and UNESCO did not move a finger until recently if it were not for the angry complaints of the *politically correct* doctor" (italics mine).

The "Black Man" issue reached the height of bad taste when a blow-up doll painted black, wearing a curly black wig with bones tied to it and an off-the-shoulder shirt with a crocodile painted on it, and carrying a spear in one hand and a shield in the other was paraded through Banyoles during Carnival. On February 10, the following headline appeared in the main Catalan-language newspaper, *Avui,* published in Barcelona. "The Carnival of Banyoles has had greater success than ever before thanks to the embalmed Black Warrior." The headline is under a photo of participants in the Banyoles carnival wearing blackface. They are shown in front of the blow-up doll under the town hall, on which a banner is tied reading: "Platform of Solidarity with the Darder Bushman." The article goes on to note that the carnival itself was dedicated to chocolate made in Banyoles.

Avui on April 9 carried an opinion column supporting at length the contention that the Darder museum, including the "Black Man" "Serves an important didactic end. The author Xvier Barral i Altet, who frequently comments on art and architecture on Avui, makes the strong claim that to consider the exhibition of the 'Black Man' a violation of human rights or a lack of respect for the dead is absurd and partial."

Another op-ed column concerning the "Black Man" was published in *Avui* on March 16, 1997. This one, close to a full page in length, was written by the well-known architect Oriol Bohigas. Sarcastic in tone and written from the perspective of the left of the political spectrum, it pretends a sense of shock at the rather "outdated" and pagan necessity to honor the dead with funerals and makes fun of the belief in immortality. Bohigas went on to suggest that the best way to achieve immortality is to have one's body displayed in a museum. He added that the Darder museum played a dignified role in the education of Banyoles's children and suggests that the exhibition serves to stimulate an interest in science, history, and literature. The article ended tongue in cheek: "Perhaps there is a means of satisfying the ingenious anti-racists. The mayor of Banyoles could present a gift to some African museum of an embalmed white and thus equilibrate the supposed wrongs."

After following the issue of the "Black Man" of Banyoles in the press for over one year, visiting the museum, where I was able to purchase a color postcard of the "Black Man," and interviewing Dr. Arcelin, I asked myself why there was this almost general blindness to the symbolism displayed in the exhibit of a stuffed African. After all, I reasoned, Catalans are highly sensitive to the role symbolism plays in the fostering and protection of a national image and identity. They are basically tolerant of differences among people who live in their midst, even as they demand that those who are different assimilate into Catalan culture via the language. They do not demand that people be monolingual in Catalan, which would be an absurdity when one considers the relationships between Catalonia and Spain. Most Catalans realize that if they were monolingual they would cut themselves off from a vast array of cultural and economic possibilities with Spain and the rest of the Castilian-speaking world.

It seems to me that the expediency displayed by this case of symbolic racism is the current crisis in Catalan. The political parties indigenous to Catalonia, with the exception of the Party Popular (PP), all carry local Catalan names, but the major left-wing party, the PSC, is allied in

Madrid with the PSOE (the socialist party of Spain). Though all parties within Catalonia, with the exception of the minority right and the extreme right wing, were in one sense Catalinist in 1977, when a million people marched in the streets of Barcelona carrying the Catalan flag and proclaiming their political autonomy, this is no longer the case. The political situation in the recent past has become more complicated as the two large Spanish national political parties, the PP on the right and the PSOE on the left, play for central power in Madrid, often at the expense of the three constitutionally recognized "historical" autonomies: Catalonia, the Basque country, and Galicia. Additionally, the leadership of the center-right controlling party in the Catalan government, the CiU, has not yet groomed the next generation for leadership and at the same time faces strong challenges in the near future from both the PP and the PSC, which is much less nationalist than the CiU. In this environment the case of the "Black Man," and indeed any controversy, no matter how symbolic, heightens already-present strains and conflicts in the local struggle for power. The fact that the battle against the exhibition of the "Black Man" was led by a single individual who, it is claimed, has not chosen to integrate himself into Catalan language and culture has created a "circle the wagon" mentality among many Catalans of all political stripes and blinded them to the racism displayed by continuing the exhibition in Banyoles. What had gone unnoticed in all this is the fact that Europeans and Americans at the turn of the century were capable of reducing native populations to objects. The example of the "Black Man" of Banyoles shows how racial insensitivity can remain a force well into modern times and how within the context of a social or economic crisis race prejudice can rise to the surface.

Finally in May of 2000, under pressure from African governments, UNESCO, and the central government, the new mayor of Banyoles, a member of the left nationalist party (the ERC), declared that the "Black Man" would be returned to Africa for burial; so he was in the fall of 2000. Let us hope that the exact copy of this unfortunate exhibit will never find its place in the Banyoles museum. The question remains: Could an incident of this type occur in the United States today? Probably not, particularly in the current climate of sensitivity to racial matters common to most public institutions such as universities and museums. But times and sensitivities change with the political winds, and what is or is not acceptable behavior is closely linked to historical circumstances. Some of our leading natural history museums still have drawers full of scalps taken in the nineteenth century by both whites and Indians. Thomas Powers reviews several books on Kit Carson, the cowboy hero, in the November 1, 2001, *New York Review of Books,* and he states: "Killing Indians and taking scalps, glorious in Cody's [and Carson's] day, are generally condemned as barbarous now, but what people say changes more quickly than what they do. The body count of Vietnamese killed by Americans in any fiscal quarter of the war in Vietnam probably would have finished off the whole Sioux nation, with maybe a smaller tribe or two thrown in, and more than one American soldier kept personal score by taking the ears of his victims."

ON THE PLUSES AND MINUSES OF MULTICULTURALISM

When I was growing up in the 1930s and 1940s the United States was touted as a "melting pot" into which immigrants would be absorbed, later emerging as full members of a united American nation. While this ideal worked with varying speed and success for whites from Europe (already established native-born Americans were more willing to grant equal status to people of northern European origin than to those whose roots were in southern Europe), an alien exclusion act passed in the 1920s and not repealed until the 1950s forbade citizenship to people of Asian origin. One need not be reminded that native Americans and blacks were also essentially excluded from the melting pot ideal.

The civil rights movement of the 1960s led to the abolition of legal segregation in the South. A new spirit of positive self-evaluation based on black pride (Black Is Beautiful) arose among members of the black community. Other groups that continued to feel the pain of economic exclusion and its social effects formed their own movements. Feminists organized, sometimes with blacks and whites in consort and sometimes separately and with different short-range goals, to challenge social and economic discrimination against women. Various Latino groups (for example, Mexicans in California, Florida, and Texas, and Puerto Ricans in New York) organized to fight discrimination and exert political and economic pressure on the majority white population. Native Americans also joined the fray by asserting their own cultural pride. They formed militant groups, the best known of which is AIM (the American Indian Movement). These organizations recruited lawyers sympathetic to the Native American cause to fight discrimination and reclaim land lost through treaty violations by federal and state governments.

Today, in spite of desegregation laws, de facto segregation continues to exist. It is propagated by the mechanism of economic exclusion (even in the face of affirmative action laws) for the vast majority of blacks in the North as well as in the South. Women too, are subject to discrimination in the market place, and minorities other than blacks face similar problems. The boiling over of group identity (not limited to minorities), saw a move away from the ideology of the melting pot toward what came to be known as "multiculturalism." Separate "racial" or ethnic groups and feminists began to lay claim to the values of their own special identities and histories within the matrix of American culture and combat what they saw as damaging stereotypes. As a major means for change these groups attacked the standard historical and cultural curricula of American universities, demanding new courses covering such topics as feminist and African American history and literature as well as philosophy courses that taught the cultural contributions of minority populations and women. In many universities this attack on "dead white men" led to the expansion of course offerings and/or the incorporation of authors who had until then been ignored in standard courses. Students and professors also lobbied for and often succeeded in obtaining new departments or programs that focused on these special interests. At my own university, Columbia, the choice was to create academic programs with appointments to be made in various traditional departments so that the new professors in these specialties would not be segregated away from the mainstream university structure. So, for example, the programs in women's studies and African American studies include faculty from literature, political science, history, sociology, and anthropology departments. The same is true for the newer programs in Latino-American studies and Asian American studies.

While I applaud the new focus on women and minorities in academic institutions it has, in some cases, had a perverse effect on the educational system. We have seen in chapter seven how Leonard Jeffries, the past chair and current professor of black studies at City College in New York, has, from within his own segregated department, been able to push an extreme Afro-centric point of view that has little historical validity. (I am among those who also criticize an extreme Eurocentric point of view, reminding my students, for example, that even in ancient times the Mediterranean was a small lake with cultural elements passing *back and forth* across the water and that land masses connected Africa and the Middle East to Eastern Europe.)

At Columbia, instead of junking our "core" humanities and literature courses the faculty has expanded the core curriculum to include courses on literature, philosophy, and social thought by minority and third-world authors. Students are expected to take a selection of these courses along with the more traditional "Contemporary Civilization" (a course that reads philosophers from the Greeks up to the modern era) and a series of humanities courses in literature, art, and music. Teachers of the core are required to assign a set of readings common to all the sections (class size is limited and several sections are offered) but in addition are free to choose other reading material to assign for their own classes.

This is a solution with which I am in complete agreement. While it is necessary to expose students to intellectual material they might not find on their own or in more traditional courses, I for one, would hate to junk the "Dead White Men," because they constitute the core of our common history and culture and because they were, in fact, great thinkers and writers. I am not a member of the postmodern camp that takes the extreme relativist position saying all ideas and all authors are of equal value. On the other hand it is important to understand why so many of the past "great" thinkers were men and to expand our horizons by reading thinkers of great value that have normally been left out of the cannon. The tendency to circle the wagons around a particular cultural group, sometimes at the expense of another, and to demand exclusive proprietary rights to a particular worldview or cultural practice is divisive and counterproductive. Multiculturalism has a positive force on American culture as a whole, and on individual groups in particular, when the particularities of a culture can be shared and appreciated by members of other cultures as well. This is what Columbia's extended core is designed to do. Let me give an example from my own experience of what I mean by the dangers of exclusivity as opposed to the advantages of sharing.

Every year at All Hallows, people who share the Mexican cultural heritage prepare a special room decorated with artifacts related to a departed loved one and consecrated to his or her memory. For a number of years a Mexican American social activist, local artist, and curator in the city of San Francisco, Rene Yañez, has asked members of the Mexican American community to create art works based on the room of the dead theme to be shown in the local Mexican American cultural center. In the early 1990s, San Francisco built the Yerba Buena Center for the Arts designed to serve the city's population, which includes a large and varied number of ethnic groups. The center, well located in the "downtown" area and across the street from the new Museum of Modern Art, consists of a large, carefully landscaped park, a theatre, and a large art gallery for temporary shows. Each of these is designed to highlight a different aspect of San Francisco's cultural variety. Shortly after the Yerba Buena Center opened, Rene Yañez was asked by its directors to prepare a "room for the dead" show. Yañez decided that it would be interesting if, this time, the show could be opened to local artists from the city's various ethnic groups, feminists, and gays, each of whom was asked to submit a project that fit the overall theme but at the same time represent their own cultural interpretation. A selection of these projects was chosen by the curator based on their artistic and cultural value. Among the many ethnic groups and feminists represented in the show were African Americans, various people of European origin, Japanese Americans, Chinese Americans, Filipino Americans, and, of course, Mexican and other Latin Americans. One woman who had survived a severe case of breast cancer chose to interpret her own brush with death. In her darkened "room" she hung an oversized hospital gown made of gauzy material on which had she printed color photographs of her own diseased cells taken during chemotherapy. The gown was lit from the inside. Transformed into an enormous lamp it was at the same time both strikingly morbid and beautiful.

My daughter, Julie Alland, is an artist specializing in assemblage (sculptural pieces of varying size made up of several parts united by a central theme). Rene Yañez asked her and a friend, Alexandria Levin, to present a project for the Yerba Buena show. Julie and Alexandria designed a room in honor of Alexandria's cousin who had died of AIDS and had not publicly revealed his homosexuality. The room's symbolism was a mix of themes borrowed from Jewish and ancient Greek culture. The centerpiece was a water-filled stone well surrounded by tree branches set into the well. At the bottom, under the water, the visitor could see a face obscured by a layer of wax. The walls of the room were hung with pictures obscured by coverings in keeping with Jewish mourning custom. At the top of the room a series of reeds hung with pomegranates delineated a roof, reminding the viewer of Persephone's descent into hell and the consequent creation of the winter season.

A room honoring a black woman who had been a seamstress and quiltmaker was a reproduction of her workshop. A Japanese woman artist used the cutting of her long braid as a symbol of mourning. The cut braid was displayed in a museum case, and the walls were hung with photographs of the cutting process. A room in honor of Holocaust victims had black walls covered with chalk marks as a numerical record of the dead. A child's wagon holding a photo album of victims was placed in the center of the space.

One of the most touching and artistically successful exhibits was in memory of those who had died of AIDS in San Francisco since the beginning of the epidemic. Over a thousand paper matches with red tips were glued to a copper base. The overall effect was one of a miniature flower garden. On the "Day without Art," which takes place every December in museums around the country to commemorate those who had died of AIDS, the exhibit was taken outside and the matches were lit as a sign of grief and respect for the victims of this terrible disease.

The show was a success. But there were dissenting voices from those who take a circle-the-wagon approach to multiculturalism. These feelings were expressed by a minority of individuals within San Francisco's large Mexican American community who were unhappy to see an important aspect of their own culture "expropriated" by others. In general, however, professional critics and visitors were impressed by the rich interplay of themes, each interpreting, in very personal ways, the common elements to be found in the main theme of the show.

The first rooms for the dead exhibition and those that followed it in successive years were a great success, fulfilling both the artistic and cultural mandate of the Yerba Buena Cultural Center and its goal of presenting a nonexclusive aesthetic of multiculturalism and an example of the value of cultural borrowing.

COUNTING AMERICANS

As the reader undoubtedly knows, every ten years the U.S. Constitution demands that a census of the entire population take place. Individuals filling out the forms have traditionally been asked to mark their racial and/or ethnic identity by choosing one answer from a short list of possible choices. This type of classification is in keeping with the notion that a person's racial (and even cultural) identity is all one piece, perhaps even genetic. When it comes to African Americans, because the American folk system of racial classification is based on the notion that "one drop of African blood puts one into that racial category," the traditional system of categorization was generally accepted, although there have always been dissenters who refuse to pick a single racial or ethnic label. Recent years have seen more and more mixed marriages of various types—between "racial" groups, between religions, and between different ethnic groups. This mixed face of the average American has come of late to be typified by the champion golfer Tiger Woods, who can and does claim African, Southeast Asian, white, and American Indian ancestry. This identity shift was officialized in the 2000 census, which, for the first time, gave individuals an expanded laundry list of racial and ethnic groups with which to identify. Additionally, people were able to check more than one category if they chose to proclaim their mixed heritage. Although such a system has merit in terms of breaking down stereotypes about so-called pure racial and ethnic groups it was seen by some as dangerous. Blacks, for example, were warned that if they did not check off "African American" as their sole racial identity, federal public assistance programs assigned by income level, but also by racial and ethnic categories, might lose significant public funding. This problem, in my opinion, provides yet another example of how complicated the playing out of racial and ethnic identity in contemporary American society is. Until we have full equality of opportunity, until the barriers of de facto racial segregation are broken down, and until racial stereotyping of the kind described in this book end, the problem will remain part of the "American dilemma" (Myrdal 1944).

BIBLIOGRAPHY

Alland, A., Jr. 1967. *Evolution and Human Behavior.* New York: Doubleday.

———. 1971. *Human Diversity.* New York: Columbia University Press.

———. 1972. *The Human Imperative.* New York: Columbia University Press.

———. 1976. Culture et Comportement. *La Research.* 57:547–556.

Altus, W. D. 1966. Birth Order and its Sequelae. *Science.* 151:44–59.

Ardrey, R. 1961. *African Genesis.* New York: Dell Publishing Co.

———. 1966. *The Territorial Imperative.* New York: Dell Publishing Co.

———. 1970. *The Social Contract.* New York: Atheneum.

Baker, P. 1960. Climate, Culture, and Evolution. In *The Process of Ongoing Human Evolution,* edited by Gabriel W. Lasker. Detroit: Wayne State University Press.

Beatty, J. 1968. Taking Issue with Lorenz on the Ute. In *Man and Aggression,* edited by Ashley Montague. New York: Oxford University Press.

Benedict, R. and G. Weltfish. 1943. *The Races of Mankind.* New York: Public Affairs Publications.

Binet, A. and T. Simon. 1916. The Development of Intelligence in Children. Translated from articles in *L'Année Psychologique* from 1905, 1908, and 1911 by Elizabeth S. Kite. Baltimore: Williams and Wilkins.

Blum, H. 1961. Does the Melanin Pigment of Human Skin Have Adaptive Value? *Quarterly Review of Biology.* 36:50–63.

Boyd, W. C. 1950. *Genetics and the Origin of Species.* Boston: Little Brown and Co.

———. 1963. Four Achievements of the Genetical Method in Physical Anthropology. *American Anthropologist.* 65:243–252.

Brace, C. L. 1964. A Nonracial Approach Towards the Understanding of Human Diversity. In *Concept of Race,* edited by Ashley Montague. Glencoe, IL: The Free Press, 103–152.

Brigham, C. C. 1930. Intelligence Tests of Immigrant Groups. *Psychological Review.* 37:158–165.

Burt, C. 1958. The Inheritance of Mental Ability. *American Psychologist.* 13:1–15.

———. 1961. Intelligence and Social Mobility. *British Journal of Statistical Psychology.* 14:3–24.

———. 1963. Is Intelligence Distributed Normally? *British Journal of Statistical Psychology.* 16:175–190.

Cavalli-Sforza, L. L. 2000. *Genes, People and Languages.* New York: North Point Press.

Cavalli-Sforza, L. L., P. Menozzi, and A. Piazza. 1994. *The History and Geography of Human Genes.* Princeton, NJ: Princeton University Press.

Cohen, R. 1969. Conceptual Styles, Culture Conflict, and Nonverbal Tests of Intelligence. *American.Anthropologist.* 71:828–856.

Coon, C. 1962. *The Origin of Races.* New York: Alfred Knopf.

———. 1965. *The Living Races of Man.* New York: Alfred Knopf.

Coon, C., S. M. Garn, and J. B. Birdsell. 1950. *Races: A Study of the Problem of Race Formation in Man.* Springfield, IL: Charles C. Thomas.

Cooper, R., and J. Zubek. 1958. Effects of Enriched and Restricted Early Environments on the Learning Ability of Bright and Dull Rats. *Canadian Journal of Psychology.* 12:159–164.

Crow, J. 1969. Genetic Theories and Influences: Comments on the Value of Diversity. *Harvard Educational Review.* 39:301–309.

Darwin, C. 1896. *The Origin of Species by Means of Natural Selection or the Preservation of Favored Races in the Struggle for Life.* (With additions and corrections from the sixth and last English edition.) New York: D. Appleton and Co. (First English edition, 1859.)

———. 1896. *The Expression of Emotions in Man and Animals.* New York: D. Appleton and Co.

———. 1896. *The Descent of Man and Selection in Relation to Sex.* (New edition, revised and expanded.) New York: D. Appleton and Co. (First English edition, 1871.)

Deutch, K. W. and T. Edsall. 1972. The Meritocracy Scare. *Society.* 9:71–79.

Diamond, M. C. 1988. *The Impact of the Environment on the Anatomy of the Brain.* New York: The Free Press.

Edelman, G. M. 1987. *The Theory of Neuronal Group Selection.* New York: Basic Books.

———. 1989. *The Remembered Present.* New York: Basic Books.

Eisenberg, L. 1972. The Human Nature of Human Nature. *Science.* 76:123.

Erlenmeyer-Kimling, L. and L. F. Jarvik. 1963. Genetics and Intelligence: A review. *Science.* 142:1477–1479.

Eysenck, H. J. 1971. (U.S. title *The I.Q. Argument.*) *Race, Intelligence, and Education.* London: Templeton Smith.

———. 1981. Science, Racism, and Sexism. *Journal of Social, Political, and Economic Studies.* 16:215–250.

Fish, J. M. 2000. What Anthropology Can Do For Psychology: Facing Physics Envy, Ethnocentrism, and a Belief in "Race." *American Anthropologist.* 102: 552–563.

Freud, S. 1961. *Civilization and Its Discontents.* New York: W. W. Norton and Co.

Fried, M. 1975. *The Notion of Tribe.* Menlo Park, CA: Cummings Publication Co.

Fuller, J. L. and J. P. Scott. 1954. Heredity and Learning Ability in Infrahuman Mammals. *Eugenics Quarterly.* 1:28–43.

Garn, S. M. 1961 and 1965. *Human Races.* Springfield, IL: Charles C. Thomas.

Geber, M. 1958. The Psychomotor Development of African Children in the First Year, and the Influence of Maternal Behavior. *Journal of Social Psychology.* 47:185–195.

Gottesman, I. I. 1968. Biogenetics of Race and Class. In *Social Class, Race, and Psychological Development,* edited by Martin Deutsch, Irwin Katz, and Arthur R. Jensen. New York: Holt, Rinehart, and Winston, 11–51.

Gould, S. J. 1981. *The Mismeasure of Man.* New York: W. W. Norton and Co.

Gould, S. J. and N. Eldredge. 1993. Punctuated Equilibrium Comes of Age. *Nature.* 366:223–227.

Gourevitch, P. 1992. The Jeffries Affair. *Commentary.* 93:34–38.

Hartley, J. and D. Rooum. 1983. Sir Cyril Burt and Typography: A Re-evaluation. *British Journal of Psychology.* 74:203–212.

Haywood, H. C. 1968. *Social-Cultural Aspects of Mental Retardation.* New York: Appleton-Century Crofts.

Hearnshaw, L. S. 1979. *Cyril Burt, Psychologist.* Ithaca, NY: Cornell University Press.

Heber, R. F., and R. B. Dever. 1968. Research on Education and Habilitation of the Mentally Retarded. In *Social-Cultural Aspects of Mental Retardation,* edited by H. C. Haywood. New York: Appleton-Century Crofts.

Herrnstein, R. 1971. I.Q. *The Atlantic Monthly.* 228:43–64.

———. 1973. *I.Q. in the Meritocracy.* Boston: Atlantic-Little Brown and Co.

———. 1990. Still an American Dilemma. *Public Interest.* 98:3–63.

Herrnstein R. and C. Murray. 1994. *The Bell Curve: Intelligence and Class Structure in American Society.* New York: The Free Press.

Hess, R. D. 1955. Controlling Cultural Influence in Mental Testing: An Experimental Test. *Journal of Educational Research.* 49:53–58.

Hiernaux, J. 1964. The Concept of Race and the Taxonomy of Mankind. In *The Concept of Race,* edited by Ashley Montague. Glencoe, IL: The Free Press.

———. 1968. *La Diversité Humaine en Afrique Subsaharienne.* Brussels: Editions de l'Institute de Sociologie Université Libre de Bruxelles.

Holden, C. 1973. The Perils of Expounding Meritocracy. *Science.* 181:36–39.

Huxley, T. H. 1910. Emancipation—Black and White. In *Huxley's Letters and Lay Sermons.* London: J. M. Dent and Sons Ltd.

Jeffries, L., Jr. 1986. Civilization or Barbarism: The Legacy of Cheikh Anta Diop. *Journal of African Civilization.* 8:146–160.

Jensen, A. J. 1968. Another Look at Culture Fair Testing. In *Western Regional Conference on Testing Problems. Proceedings.* 1968. Princeton, NJ: Educational Testing Service.

———. 1969a. How Much Can We Boost IQ and Scholastic Achievement? *Harvard Educational Review.* 39:1–123.

———. 1969b. Reducing the Heredity-Environment Uncertainty: Comments on the Value of Diversity. In *Environment, Heredity, and Intelligence.* Reprint series no. 2. Compiled for *Harvard Educational Review,* 209–243.

———. 1972a. *Genetics and Education.* London: Methuen.

———. 1974a. How Biased Are Culture-Loaded Tests? *Genetic Psychology Monographs.* 90:185–244.

———. 1974b. Cumulative Deficit: A Testable Hypothesis? *Developmental Psychology.* 10:996–1019.

———. 1977. Cumulative Deficit in IQ of Blacks in the Rural South. *Developmental Psychology.* 13:184–191.

———. 1978a. Sir Cyril Burt in Perspective. *American Psychologist.* May: 499–503.

———. 1978b. Genetic and Behavioral Effects of Nonrandom Mating. In *Human Variation: Biopsychology of Age, Race, and Sex,* edited by R. T. Osborne, C. E. Noble, and N. Weyl. New York: Academic Press, 5–105.

———. 1980. *Bias in Mental Testing.* New York: Free Press.

———. 1984a. The Black-White Differences on the K-ABC: Implications for Future Tests. *Journal of Special Education.* 18:377–408.

———. 1984b. Test Bias: Concepts and Criticisms. In *Perspectives on Bias in Mental Testing,* edited by C. R. Reynolds and R. T. Brown. New York: Plenum Press, 507–586.

———. 1985. The Nature of the Black-White Difference on Various Psychometric Tests: Spearman's Hypothesis. *The Behavioral and Brain Sciences.* 8:193–258.

———. 1986. g: Artifact or Reality. *Journal of Vocational Behavior.* 29:301–331.

———. 1987a. Continuing Commentary on the Nature of the Black-White Difference on Various Psychometric Tests: Spearman's Hypothesis. *The Behavioral and Brain Sciences.* 10:507–537.

———. 1987b. The g beyond Factor Analysis. In *The Influence of Cognitive Psychology on Testing,* edited by R. Royce et al. Hillsdale, NJ: Lawrence Erlbaum Associates, 87–142.

———. 1989. Raising I.Q. without Raising g? A Review of the Milwaukee Project: Preventing Mental Retardation in Children at Risk. *Developmental Review.* 9:234–258.

———. 1990. Speed of Information Processing in Calculating Prodigy. *Intelligence.* 14:259–274.

———. (1992). Spearman's Hypothesis: Methodology and Evidence. *Multivariate Behavioral Research.* 27:225–233.

———. 1993a. Psychometric g and Achievement. In *Perspectives on Educational Testing,* edited by D. R. Gifford. Boston: Kluwer Academic Publishers, 117–227.

———. 1993b. Spearman's Hypothesis Tested with Chronometric Information-Processing Tasks. *Intelligence.* 17:44–77.

Jolly, A. 1966. *Lemur Behavior.* Chicago: University of Chicago Press.

Kamin, L. J. 1974. *The Science and Politics of I.Q.* New York: John Wiley and Co.

Katz, I. 1968. Some Motivational Determinants of Racial Differences in Intellectual Achievement. In *Science and the Concept of Race,* edited by M. Mead, T. Dobzhansky, E. Tobach, and R. E. Light. New York: Columbia University Press, 132–148.

Kendall, L. M., M. A. Verster, and J. M. Von Mollendorff. 1988. Test Performances of Blacks in Southern Africa. In *Human Abilities in Cultural Context,* edited by S. H. Irvine and J. W. Berry. Cambridge: Cambridge University Press. 299–339.

Kincheloe, J. L., S. R. Steinberg, and A. D. Gresson III. 1996. *Measured Lies: The Bell Curve Examined.* New York: St. Martin's Press.

Kuhl, S. 1994. *The Nazi Connection.* New York: Oxford University Press.

Leeds, A. 1971. The Concept of the Culture of Poverty: Conceptual, Logical and Empirical Problems with Perspectives from Brazil and Peru. In *Culture of Poverty,* edited by E. B. Leacock. New York: Simon and Schuster, 226–284.

Lelyveld, J. 2001. *How Race Is Lived in America: Pulling Together, Pulling Apart.* New York: Henry Holt and Co.

Levin, M. E. 1981. Equality of Opportunity. *Philosophical Quarterly.* 31:110–125.

———. 1984a. Why Homosexuality Is Abnormal. *The Monist.* 67:251–283.

———. 1984b. Comparable Worth: The Feminist Road to Socialism. *Commentary.* 78:13–19.

———. 1986. Feminism, Stage Three. *Commentary.* 82:27–31.

———. 1991. Race Differences: An Overview. *Journal of Social, Political, and Economic Studies.* 16:195–216.

———. 1997. *Why Race Matters: Race Differences and What They Mean.* Westport, CT: Praeger.

Lewis, O. 1969. *Five Families.* New York: Basic Books.

Lewontin, R. C. 1972. The Apportionment of Human Diversity. In *Evolutionary Biology,* edited by M. K. Hecht and W. S. Steere. Vol. 6. New York: Plenum.

———. 1995. *Human Diversity.* New York: Scientific American Library. New York: W. H. Freeman and Co.

Livingstone, F. 1958. Anthropological Implication of Sickle Cell Gene in West Africa. *American Anthropologist.* 60:533–562.

Lorenz, K. 1967. *On Aggression.* New York: Bantam Books.

———.1973. *Les Huit Péchés Capitaux de Notre Civilization.* Paris: Flammarion. English edition 1974.

McAskie, M. 1978. Carelessness or Fraud in Sir Cyril Burt's Kinship Data? A Critique of Jensen's Analysis. *American Psychologist.* May: 496–498.

McCombs, R. C. and J. Gay. 1988. Effects of Race, Class, and I.Q. Information on Judgments of Parochial School Teachers. *The Journal of Social Psychology.* 128:647–652.

Myrdal, G. 1944. *An American Dilemma: The Negro Problem and Modern Democracy.* New York: Harper and Brothers.

Putnam, C. 1961. *Race and Reason: A Yankee View.* Washington, D.C.: Public Affairs Press.

Ramey, C. T., D. MacPhee, and K. O. Yeates. 1982. Preventing Developmental Retardation: A General Systems Model. In *How and How Much Can Intelligence Be Increased,* edited by D. K. Detterman, and R. J. Sternberg. Norwood, NJ: Ablex Publishing Co.

———. 1992. High-Risk Children and IQ: Altering Intergenerational Patterns. *Intelligence.* 16:239–256.

Ricker, J. P. and J. Hirsch. 1988. Genetic Changes Occurring over 500 Generations in Lines of Drosophila Melanogaster Selected Divergently for Geotaxis. *Behavior Genetics* 18, no. 1:13–25.

Rosenthal, R., and L. F. Jacobson. 1969. *Pygmalion in the Classroom: Teacher's Expectations of Pupils' Intellectual Development.* New York: Holt, Rhinehart and Winston.

———. 1968. Teacher Expectation for the Disadvantaged. *Scientific American.* 218:19–23.

Rushton, J. P. 1990. Race, Brain Size, and Intelligence: A Reply to Chernovsky. *Psychological Reports.* 66:659–666.

———. 1992a. Cranial Capacity Related to Sex, Rank, and Race in a Stratified Sample of 6,235 U.S. Military Personnel. *Intelligence.* 16:401–413.

———. 1992b. Evidence for Race and Sex Differences in Cranial Capacity From International Labour Office Data. Research Bulletin 712 (manuscript).

———. 1995. *Race, Evolution, and Behavior: A Life History Perspective.* New Brunswick, NJ: Transaction Publishers. (Abridged editions 1999 and 2000.)

Scott, J. P. and J. L. Fuller. 1965. *Genetics and the Social Behavior of the Dog.* Chicago: University of Chicago Press.

Shockley, W. 1972. Dysgenics, Geneticity, Raceology: A Challenge to the Intellectual Responsibility of Educators. *Phi Delta Kappa.* 297–307.

———. 1972. A Debate Challenge: Geneticity is 80% for White Identical Twins' I.Q.s. *Phi Delta Kappa.* 415–419.

Shuey, A. 1966. *The Testing of Negro Intelligence.* New York: Social Science Press.

Spearman, C. 1904. General Intelligence Objectively Determined and Measured. *American Journal of Psychology.* 15:201–293.

Stewart, O. C. 1968. Lorenz/Margolin on the Ute. In *Man and Aggression,* edited by Ashley Montague. New York: Oxford University Press.

Terman, L. M. 1937. *The Measurement of Intelligence.* Boston: Houghton Mifflin.

Thorndike, E. L. 1927. *The Measurement of Intelligence.* New York: Bureau of Publications, Teacher's College, Columbia University.

Trotman, F. K. 1976. Race, I.Q., and the Middle Class. Ph.D. diss., Graduate School of Arts and Sciences, Columbia University.

Tryon, R. C. 1940. Genetic Differences In Maze-Learning Ability in Rats. *Yearbook of the National Society for Studies in Education.* 39, no. 1:111–119.

Wade, N. 1976. I.Q. and Heredity: Suspicion of Fraud Beclouds Classic Experiment. *Science.* 194:916–919.

Wheeler, L. R. 1942. A Comparative Study of the Intelligence of East Tennessee Mountain Children. *Journal of Educational Psychology.* 33:321–334.

Wilson, E. O. 1975. *Sociobiology.* Cambridge, MA: Harvard University Press.

INDEX